S0-BLM-099

Quantitative Analysis using Chromatographic Techniques

SEPARATION SCIENCE SERIES

Editors: Raymond P W Scott, Colin Simpson and Kenneth Ogan

Quantitative Analysis using Chromatographic Techniques

Edited by **Elena Katz**

Quantitative Analysis using Chromatographic Techniques

Edited by

Elena Katz
Perkin Elmer, USA

JOHN WILEY & SONS
Chichester · New York · Brisbane · Toronto · Singapore

Library of Congress Cataloging in Publication Data:

Quantitative analysis using chromatographic techniques.
(Separation science series)
Includes bibliographies and index.
1. Chromatographic analysis. 2. Chemistry, Analytic—Quantitative. I. Katz, Elena. II. Series.
QD117.C5Q36 1987 545'.89 86-26715

ISBN 0 471 91406 1

British Library Cataloguing in Publication Data:

Quantitative analysis using chromatographic techniques.—(Separation science series)
1. Chromatographic analysis 2. Chemistry, Analytic—Quantitative
I. Katz, Elena II. Series
545.89 QD117.C5

ISBN 0 471 91406 1

Printed and bound in Great Britain

CONTRIBUTORS

Eli Grushka, Department of Inorganic and Analytical Chemistry, The Hebrew University, Jerusalem, Israel

Eric C. Jensen, Lilly Research Laboratories, Eli Lilly and Company, Indianapolis, IN 46285

Salwa Khatib, Department of Chemistry, Wayne State University, Detroit, MI 48202

Shulamit Levin, Department of Inorganic and Analytical Chemistry, The Hebrew University, Jerusalem, Israel

G. Raymond Miller, Alcon Laboratories, Inc., 6201 South Freeway, Fort Worth TX 76101

Kenneth Ogan, The Perkin-Elmer Corporation, 761 Main Ave., Norwalk, CT 06859-111

Colin F. Poole, Department of Chemistry, Wayne State University, Detroit, MI 48202

Charles E. Reese, Department of Chemistry, Birkbeck College, University of London, Malet Street, London WC1E 7HX, England

Raymond P.W. Scott, The Perkin-Elmer Corporation, 761 Main Ave, Norwalk, CT 06859-0089

Peter C. Uden, Department of Chemistry, University of Massachusetts, Amherst, MA 01003

CONTENTS

CHAPTER 1. INTRODUCTION TO THE CHROMATOGRAPHIC PROCESS 1
Raymond P.W. Scott

PRINCIPLES OF A CHROMATOGRAPHIC SEPARATION 2
Factors That Control Selectivity 3
Ionic Forces 4
Polar Forces 4
Dispersive Forces 4
METHODS OF DEVELOPMENT IN CHROMATOGRAPHY 5
Frontal Analysis 5
Elution Development 8
Displacement Development 9
Methods of Elution Development 11
Flow Programming 12
Temperature Programming 12
Gradient Elution 12
THE FUNCTION OF THE CHROMATOGRAPHIC COLUMN/THIN-LAYER PLATE 14
Dispersion in a Packed Column 14
The Multipath Process 16
Longitudinal Diffusion Process 16
Dispersion Due to Resistance to Mass Transfer 17
Extra-Column Dispersion 19
THE BASIC CHROMATOGRAPHIC SYSTEM 20
The Gas Chromatograph 20
The Liquid Chromatograph 21
The Thin-Layer Plate System 23
THE ROLE OF THE DETECTOR 23
Linearity 24
Sensitivity 24
Detector Dispersion 26

SYNOPSIS 26
References 27

CHAPTER 2. DETECTION IN QUANTITATIVE LIQUID CHROMATOGRAPHY 31
Kenneth Ogan

DETECTOR PERFORMANCE 31
Linearity 31
Sensitivity 32
System Noise 37
DATA HANDLING 38
Peak Heights and Peak Areas 39
Signal Digitization 47
Sources of Error 44
Gaussian Peaks 45
Non-Gaussian Peaks 46
Baseline Drift 49
Incomplete Resolution 50
PEAK IDENTIFICATION 54
Multiple Detectors 54
Multi-Channel Detectors 57
SYNOPSIS 59
References 60

CHAPTER 3. QUANTITATIVE ANALYSIS BY LIQUID CHROMATOGRAPHY 63
Raymond P.W. Scott

OPERATING PROCEDURES FOR QUANTITATIVE LIQUID CHROMATOGRAPHY 65
Choice of Detectors 70
Sample Preparation 74
Injection Procedures 79
Operating Conditions 80
QUANTITATIVE ANALYSIS 85
The Internal Standard Method 86
The External Standard Method 87
The Normalization Method 88
THE PRECISION OF MODERN LC ANALYSES 89
SYNOPSIS 95

References 97

CHAPTER 4. DETECTION IN QUANTITATIVE GAS CHROMATOGRAPHY 99
Peter C. Uden

CLASSIFICATION OF GC DETECTORS 100
Mass Sensitive Detectors 102
Concentration Sensitive Detectors 103
Universal Detection 104
Selective Detection 105
Specific Detection 105
CLASSIFICATION BASED ON THE DETECTION PROCESS 105
Ionization Detection 105
Bulk Property Detectors 106
Spectroscopic Detectors 107
Electrochemical Detectors 108
Reaction Based Detectors 108
GENERAL QUANTITATIVE ASPECTS OF GC DETECTORS 109
Baseline Stability 109
Signal-to-Noise Ratio 110
Minimum Detectable Level 112
Linear Dynamic Range 113
Standardization 116
DETECTION SYSTEMS AND THE INTEGRAL GAS CHROMATOGRAPH 117
Packed and Open Tubular Columns 118
Injection and Sample Introduction 120
Effect of Dead Volume and Band Broadening on Detection 121
Multiple Detectors 121
GAS CHROMATOGRAPHY DETECTORS-CHARACTERISTICS AND COMPARISONS 123
Established, Historical and Developmental Detectors 123
The Thermal Conductivity Detector (TCD) 123
The Flame Ionization Detector 127
Modifications to the FID Operation 129
The Alkali FID (AFID) 130

The Photoionization Detector (PID) 131
The Electron Capture Detector (ECD) 133
The Mass Spectrometer Detector (MSD) 134
The Infrared Spectrometer Detector 136
The Flame Photometric Detector (FPD) 137
The Electrolytic (Hall) Detector (HECD) 138
SPECIAL DETECTING SYSTEMS 139
The Gas Density Balance Detector (GDBD) 139
Modified Thermionic Ionization Detectors (TID) 140
Inert Gas (Helium and Argon) and Universal Ionization Detectors 141
Atomic Spectrometer Detectors 142
Flame Emission Detectors (FED) 143
Atomic Absorption Detector (AAD) 143
Atomic Emission Spectrometer (AED) 143
The Microwave Induced Plasma Detector (MIP) 144
The DC Argon Detector (DCP) 147
The Inductively Coupled Argon Plasma Detector (ICP) 147
The UV Absorption Detector 147
The Ultrasonic Detector (USD) 148
Chemiluminescent Detectors 148
SYNOPSIS 149
References 152

CHAPTER 5. QUANTITATIVE ANALYSIS BY GAS CHROMATOGRAPHY 157
Charles E. Reese and Raymond P.W. Scott

SAMPLE PREPARATION 157
SAMPLE INTRODUCTION 159
Syringe Injection 159
Injection Valves 161
Solid Sample Injection 163
Split-Flow Injection 164
Splitless Injection 166
Retention Gap Method 168
Solute Focussing Method 170

RESOLUTION 172
CHOICE OF DETECTORS 175
METHODS OF PEAK AREA MEASUREMENT 177
Height X Width at Half-Height 178
The Triangulation Method 179
The Planimeter Method 180
The Cut and Weigh Method 181
Peak Height X Retention 182
PEAK HEIGHT MEASUREMENT 182
ELECTRONIC INTEGRATION 183
Processing the Data for Quantitative Analysis 184
SYNOPSIS 187
References 189

CHAPTER 6. QUANTITATIVE THIN-LAYER CHROMATOGRAPHY 193
Colin F. Poole and Salwa Khatib

INTRODUCTION 193
SAMPLE SEPARATION 196
Chromatographic Performance of a Thin-Layer Plate 196
Continuous Sample Development 201
Multiple Sample Development 202
Two-Dimensional Sample Development 209
Circular and Anticircular Development 210
SAMPLE APPLICATION 211
Solvent Selection 212
Sample Application Techniques 213
SCANNING DENSITOMETRY 218
Detection Methods 218
Instrumentation for Scanning Densitometry 220
Performance of a Scanning Densitometer 223
Sensitivity 225
QUALITATIVE SAMPLE IDENTIFICATION BY SCANNING DENSITOMETRY 233
THEORETICAL CONSIDERATIONS FOR QUANTITATIVE ANALYSIS 236

PRACTICAL CONSIDERATIONS FOR QUANTITATIVE ANALYSIS 238
Calibration Procedures in the Absorption Mode 238
Graphical Linearization Techniques 240
Electronic Linearization Techniques 243
Calibration Procedures in the Fluorescence Mode 245
Sources of Error in Scanning Densitometry 249
THE USE OF COMPUTERS IN SCANNING DENSITOMETRY 251
PHYSICAL AND CHEMICAL METHODS USED TO ENHANCE DETECTABILITY 253
LESS COMMON METHODS OF SAMPLE IDENTIFICATION AND QUANTITATION 259
SYNOPSIS 261
References 263

CHAPTER 7. CHROMATOGRAPHY AS A QUANTITATIVE TOOL IN PHARMACEUTICAL ANALYSIS 271
Eric C. Jensen

INTRODUCTION 271
SEPARATION DEVELOPMENT 272
SAMPLE HANDLING TECHNIQUES 281
FACTORS DETERMINING THE ACCURACY AND PRECISION OF QUANTITATIVE ANALYSIS 288
Linearity 290
Precision 290
Recovery 291
Specificity 292
Detection Limit 292
Stability 293
Matrix Effects 293
System Suitability 293
EXAMPLES 294
SYNOPSIS 306
References 307

CHAPTER 8. IS AUTOMATION THE FUTURE OF QUANTITATIVE CHROMATOGRAPHY? 309
G. Raymond Miller

THE NEED FOR AUTOMATED ANALYSIS 309
Sample Preparation 313
Sample Introduction 316
Chromatographic Optimization 317
Detection 318
Fraction Collection 319
Data Analysis 320
AUTOSAMPLERS 320
Sample Selection 321
LC Autosamplers 322
GC Autosamplers 326
TLC Autosamplers 327
COMPUTERS 328
Data Acquisition 328
Data Processing 330
Post Processing 332
Control 332
Interactive Control 333
Interactive Problem Solving 336
Archiving 337
Communications 338
Artificial Intelligence 339
Retention Indicies 341
ROBOTICS 341
THE ROLE OF THE CHEMIST IN THE AUTOMATED LABORATORY 350
SYNOPSIS 351
References 355

CHAPTER 9. PHYSICO-CHEMICAL INFORMATION FROM PEAK SHAPE AND WIDTH IN LIQUID CHROMATOGRAPHY 359
Eli Grushka and Shulamit Levin

INTRODUCTION 359

DETERMINATION OF LIQUID DIFFUSION COEFFICIENTS BY THE CHROMATOGRAPHIC BROADENING TECHNIQUES 361
Theoretical Approach 361
Requirements for the Experimental System 364
Sources of Extra-Column Broadening Effects 365
Additional Sources of Error in the Measurement of Diffusion Coefficients 368
Recent Applications of the Method 371
ADSORPTION ISOTHERMS 374
The Chromatographic Peak Shape and Its Dependence on the Adsorption Isotherms 379
The Langmuir Type Isotherm 383
The Freundlich Type Isotherms 385
Adsorption Isotherm and Void Volume Determinations 387
Experimental Measurements of Isotherms 390
Static Techniques 391
Dynamic Techniques - Chromatographic Methods 392
Frontal Analysis Breakthrough Curves 392
Frontal Analysis by Characteristic Point (FACP) 393
Other methods 395
Uses of Isotherms in Liquid Chromatography 397
Solute Distribution Coefficients 397
Elucidation of Retention Process 397
Ion Pair Chromatography 398
Adsorption Chromatography 399
Ligand Exchange Chromatography 399
SYNOPSIS 400
References 402

LIST OF SYMBOLS 411

INDEX 415

PREFACE

Precision and accuracy, combined with the unique separating capability, speed and simplicity of chromatography, have made it a very powerful modern analytical tool. Although many comprehensive books and reviews have been devoted to the various aspects of chromatography, it is thought that a general text covering the quantitative principles of chromatography methods is a timely and indeed much needed, adjunct to the chromatography literature.

This book attempts to encompass all the important aspects of quantitative chromatography analyses. The first chapter introduces the essential principles of a chromatographic separation and describes the basic chromatographic systems, together with the important characteristics of detecting devices. Three general chapters (3,5,6) are devoted to quantitative features of liquid chromatography, gas chromatography and thin-layer chromatography. The chapter on the latter method also includes a discussion on TLC detection methods together with the instrumentation required. Chapter 2 discusses detection in quantitative liquid chromatography and covers essential characteristics of detector performance with emphasis on data handling procedures. Chapter 4 discusses the broad range of detectors utilized in gas chromatography. Chapter 7 gives examples of the use of chromatographic techniques in the pharmaceutical industry and Chapter 8 presents a general overview of the automation procedures utilized in the modern chromatographic laboratory. And the last chapter is addressed to those readers who are interested in the use of chromatographic techniques in physical chemistry measurements.

It is hoped that the information contained in the book will be useful to both beginners in the field and for experienced analysts with special chromatographic needs.

Finally, the editor would like to thank the authors and the publisher for the cooperation they have provided, and her colleagues, in particular, Miss Carol Halfmann, for their help. The editor is much indebted to Dr. Raymond P.W. Scott whose expertise, advice and support have been always invaluable. And very special appreciation goes to Mrs. Carol Nash whose patience and skills in the production of the camera-ready copy made this book possible.

Quantitative Analysis using
Chromatographic Techniques
Edited by Elena Katz

Chapter 1

INTRODUCTION TO THE CHROMATOGRAPHIC PROCESS

Raymond P. W. Scott

The popularity of chromatography as a quantitative analytical technique results from its duality of function. Chromatography is primarily a separation process, but if a detector of known response is employed to monitor the column eluent, then a quantitative estimation of any individual component that is discretely eluted can also be obtained. Thus, in contrast to other analytical methods, the procedure needs not be specific to a given component or component class and can quantitatively estimate the proportion of each compound present in the mixture without prior treatment.

The use of chromatography as a separation technique was first identified by the Russian botanist Tswett who published his findings in 1901 (1). His observations were helped by the fact that the substances that were of interest were colored and, therefore, the separation he obtained could be observed by the naked eye. Despite the exciting possibilities that Tswett's observation evoked, little work was carried out until 1931 when Kuhn and Lederer (2) repeated Tswett's experiments using alumina and calcium carbonate as adsorbents to again separate plant pigments. These techniques were basically liquid-solid adsorption chromatography (LSC) and it wasn't until the late 1930's that Reichstein (3) with the aid of more sophisticated detection techniques than that of the human eye introduced elution chromatography. In 1941 Martin and Synge (4) introduced liquid-liquid partition chromatography (LLC) and in that publication Martin and Synge also suggested the possibilities of gas chromatography (GC). Their suggestions were not taken up, however, and even the development of liquid chromatography at that time was very slow and sporadic. In the early 1950's James and Martin started work on the the original suggestion of Martin and Synge and developed the first gas chromatograph (5). Between 1954 and 1962 gas liquid chromatography (GLC) developed at a remarkable rate and by 1962 became a mature analytical technique. After 1962, however, the development of GC almost stopped and those improvements that did occur (except for one or two notable exceptions) were relatively superficial. The rapid development of GC in the late 1950's and early 1960's provoked further development of LC which, during the period of GC development, had become the Cinderella of the chromatographic techniques.

The renaissance of LC started in the late 1960's and by the early 1980's, LC had also become an established analytical technique used generally for the separation and analysis of a wide range of complex mixtures including many that had previously been analyzed by GC. Some of the more recent developments in LC may well provoke further synergistic developments in GC but further progress in LC appears at this time to be relatively slow just as it has been with GC over the last 2 decades. The need to separate the complex mixtures of substances that are of interest in the field of biotechnology may, perhaps, provide the necessary impetus to stimulate further developments in LC in the near future, but this remains to be seen.

PRINCIPLES OF A CHROMATOGRAPHIC SEPARATION

The chromatographic process has been classically defined as a separation that is achieved by distributing the solute mixture between two phases, a mobile phase and a stationary phase. Those solutes that are preferentially distributed in the stationary phase remain in the column longer than those that are preferentially distributed in the mobile phase. Consequently, the individual solutes will be eluted from the column in the order of their increasing distribution coefficients with respect to the stationary phase. This definition is a little trite as, although it introduces the essential concept of a stationary and a mobile phase, the method of retention is obscured by the term distribution. In fact, the solutes are distributed between the two phases to different extents because the molecular interactive forces between the solute molecules and those of the two phases are different for each individual solute. For those solutes distributed preferentially in the stationary phase, the forces between the solute molecules and stationary phase molecules are much greater than the forces between solute molecules and those of the mobile phase. Conversely, solutes distributed preferentially in the mobile phase exhibit greater forces between them and the mobile phase than they do with the stationary phase. As a consequence, to control the retention of a solute and thus the chromatographic selectivity, different molecular forces have to be exploited in the two phases, the choice of which will depend on the nature of the solute molecules.

Factors that Control Selectivity

It has already been stated that solutes distributed preferentially in the stationary phase are selectively retained and thus in GC, where there is little or no interaction with the mobile phase (a gas), the nature of the stationary phase will exclusively control the selectivity of the chromatographic system. In LC, as there are interactions between the solute and both phases, selectivity can be changed by altering the nature of either or both phases. The selectivity of either phase can be changed by exploiting the different kinds of molecular forces that can exist between the solute and phase molecules. There are basically three types of molecular forces that control solute retention.

1. Ionic Forces

Ionic forces result from permanent electrical charges that exist in the molecules, for example in ionic materials such as salts. It follows that to retain anionic materials in a chromatographic system, the stationary phase should contain cations and conversely, if cationic materials are required to be retained in a column then the stationary phase should be anionic in nature.

2. Polar Forces

Polar forces between molecules arise from permanent or induced electrical dipoles existing in the molecules that are interacting and do not result from permanent electrical charges. Examples of polar solutes would be substances having permanent dipoles, e.g. alcohols, ketones and aldehydes. Examples of polarizable substances would be benzene or toluene. Again, it follows that to selectively retain polar materials in a given chromatographic system, the stationary phase should consist of polar molecules such as hydroxy compounds or at least contain polarizable groups such as materials having aromatic nuclei.

3. Dispersive Forces

Dispersive forces are also electrical in nature, but do not arise from either a net electrical charge on the molecules or induced, or permanent dipoles. Typical dispersive interactions are those associated with hydrocarbon chains. Normal heptane is a liquid and not a gas due to the fact that dispersive interactions between each molecule in n-heptane are sufficiently strong to maintain the substance in liquid form. Consequently, to retain a solute on the basis of dispersive interactions in gas chromatography, the stationary phase could be a high boiling hydrocarbon such as Apiezon Grease and alternatively, in liquid chromatography a hydrocarbon bonded phase.

The three basic types of molecular interactions can be sub-divided further, but for the purpose of this book the division into the three basic groups is sufficient.

METHODS OF DEVELOPMENT IN CHROMATOGRAPHY

There are three basic methods of chromatographic development: displacement development, frontal analysis, and elution development, although displacement analysis might be considered a special case of elution chromatography. Of the three methods, elution chromatography is by far the most used method of development. Employing elution development, complete separation can be achieved and discrete bands of each solute obtained from the chromatographic system which is essential for accurate quantitative analysis.

Frontal Analysis

There are two important ways in which frontal analysis differs from elution development. First, in frontal analysis the sample is continuously fed onto the column during development, either as a pure sample or as a solution in the mobile phase, whereas in elution development a discrete sample is placed on the column and the chromatogram subsequently developed. Second, elution development can, under the right circumstances, be made to completely separate all the components of a mixture, whereas in frontal analysis only part of the first compound is eluted in a relatively pure state, each subsequent component being mixed with those previously eluted.

Consider a three-component mixture containing equal quantities of each component fed continuously onto a column; because of the forces between solute and stationary phase, each solute will be retained to a different extent as it comes into equilibrium with the stationary phase while passing through the column. The first component to elute will be that which is held least strongly in the stationary phase, then the second component will elute but in conjunction with the first component, and finally, the most strongly held of the three will elute in conjunction with the first and second components. Subsequently, there will be no change in concentration of solute in the mobile phase and the concentration of the respective solutes will be the same as the feed mixture. The concentration profile resulting from frontal analysis, when operated under ideal conditions, is shown in Figure 1a. The continuous curve shows the total concentration of solute in the eluent, plotted against volume of mobile phase passed through the column, and the dotted curves represent a similar concentration profile but for each individual component. The ideal nature of this separation is depicted by the vertical steps of each concentration profile as the components elute from the column. In practice, sharp fronts are not obtained from concentration profiles due to the various spreading processes that occur in a column which cause the front of each solute profile to be diffused. The effect of this is shown as the continuous curve in Figure 1b. The dotted curves again represent the diffuse front of each individual component. It can also be seen from the diagram that even the first component to be eluted may contain traces of the other two components due to the diffuse fronts.

Frontal analysis was employed as a development procedure in the early stages of chromatography and before detection procedures were fully effective; it is not often used today, and certainly not for quantitative analysis. The reason for this is that no individual component is completely separated from the others in the mixture, except possibly the first component and even this may contain significant quantities of the other materials. If, after the complete profile of the frontal analysis has been

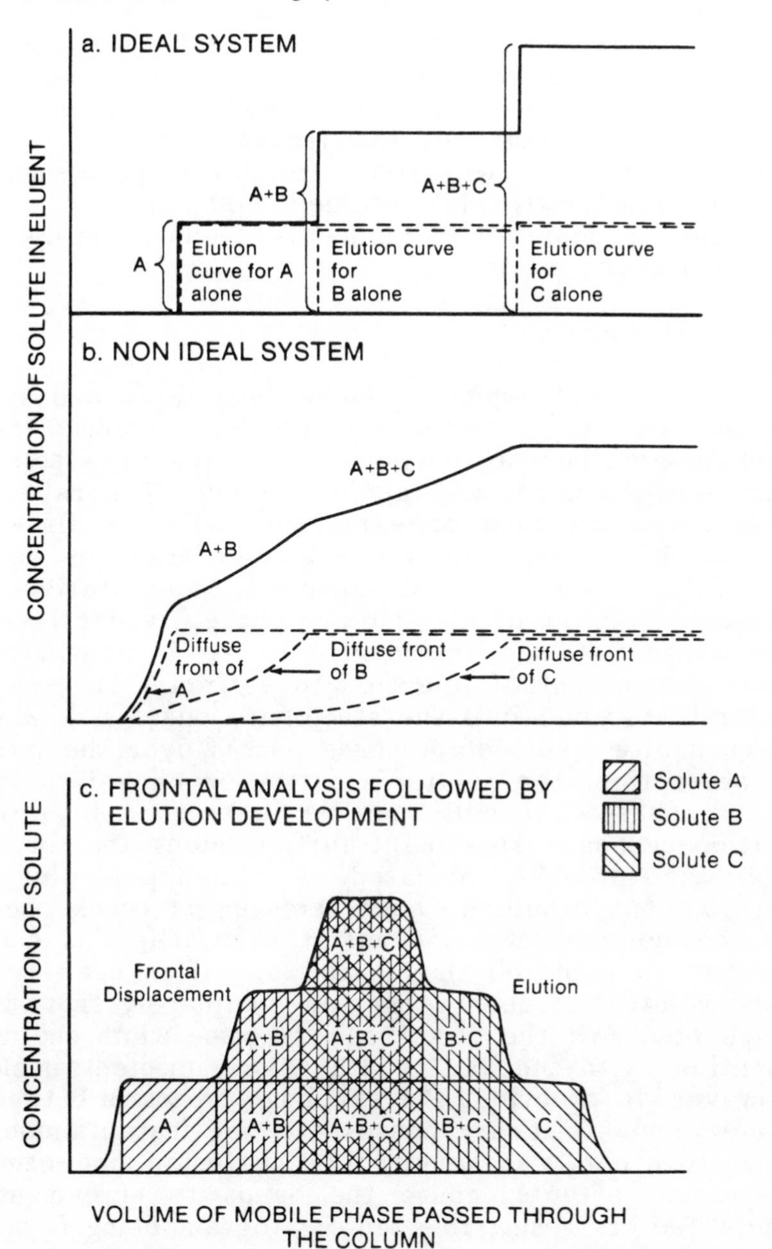

Figure 1. Concentration profile for the frontal displacement of a mixture of three components.

obtained, the sample feed is replaced by pure mobile phase, then the frontal analysis becomes elution development of a very large sample. The elution curve is the reverse of the frontal analysis curve, and the final step will give a relatively pure sample of the most strongly retained component. This situation is depicted in Figure 1c. Today, the use of frontal analysis is rare.

Elution Development

Elution development can be best described as a series of extraction-absorption processes that operate continuously during the passage of a solute band down a chromatographic column. The rate of absorption and rate of extraction will, as already discussed, depend upon the relative forces between the solute molecule and those of the stationary phase and those of the mobile phase, respectively. The absorption and extraction process is continuous from the instant of injection to elution. The solute is first absorbed into the stationary phase and, when fresh solute-free mobile phase passes over the point of absorption, the solute is extracted back into the mobile phase. It will then be reabsorbed into the stationary phase at a point further along the column and the process repeated. For ideal elution development, where no band spreading processes occur in the column, the concentration profiles of each solute is depicted in Figure 2a. It is seen that each solute has been separated completely from its neighbors. All the bands have the same width and are identical with the band width of the injected sample. However, in practice, the exchange processes between mobile and stationary phases are not instantaneous; this, together with longitudinal diffusion and other dispersive effects, cause the bands to spread and these factors result in a chromatogram being formed having the shape shown in Figure 2b. It is seen that a typical Gaussian elution curve is produced for each solute and that the extent of band-spreading increases with the extent to which the solute is retained in the column.

Elution development is by far the most effective method of development used in chromatography and is virtually the only method employed in quantitative analysis.

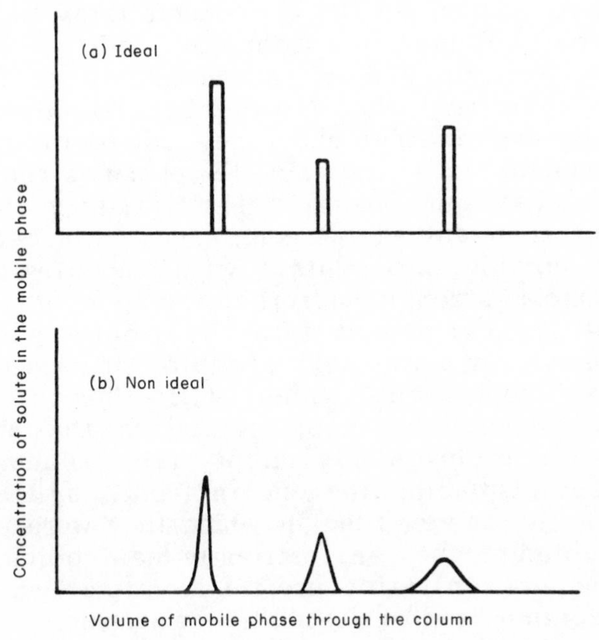

Figure 2. Elution Development of a Three-Component Mixture.

Displacement Development

Displacement development depends on the competition between solutes for the active sites of the adsorbent and is only really effective in separating very strongly adsorbed materials. In displacement development all the substances in the sample will be held on the stationary phase so strongly that they cannot be eluted by the mobile phase; they can, nevertheless, be displaced by substances that are held on the surface by stronger forces. However, there will be competition between individual solutes and, when the sample is placed on the column, all the immediately available active sites of the adsorbent will be occupied by the most strongly held component. As the band of sample is moved down the column, the next available sites will be occupied by the next most strongly retained component. Thus, all the components array themselves along the column in order of their adsorption strength, or, in effect, in order of the forces between them and the adsorbent. To develop the chromatogram, another substance called the *displacer* is introduced into the mobile phase stream; the displacer has an even higher affinity for the adsorbent than any of the components to be separated. Thus, on coming into contact with the sites occupied by the most strongly adsorbed component, it will displace this component into the mobile phase and thus move onto the next group of sites occupied by the next component which will then itself be displaced. Thus, the displacer drives the adsorbed components progressively along the column, each component displacing the one in front, until they are eluted in the same order in which they were adsorbed on the column; the least strongly held being eluted first. The concentration profile monitored at the end of the column from displacement development will take the form shown in Figure 3. The order in which the solutes emerge will characterize the individual components, and the length of the band, not the height, will be proportional to the concentration of any individual component. Because the capacity of the adsorbent may vary for each type of molecule adsorbed, the proportionality constant may not be the same for each substance. If the device used to monitor the concentration of solutes in the eluent has been made to have a response that is different for the individual components, then the height of the steps would also give a guide to the identity of the solute. For example, if a UV absorption device was used to detect the solutes, then high steps would indicate the presence of substances having a high absorption coefficient in the UV.

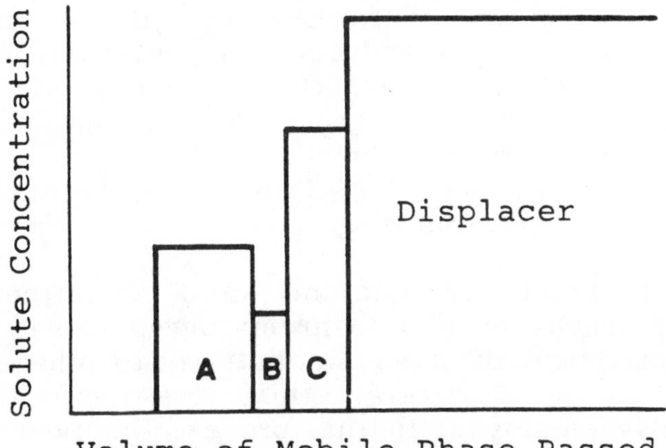

Figure 3. Displacement Development of a Three-Component Mixture.

Displacement chromatography has very limited applications as a separation technique and is only very rarely used in quantitative analysis. The reason for this is that each of the substances to be separated must have different response characteristic to the monitoring device, and furthermore, the process does not produce discrete local concentrations of individual components.

Methods of Elution Development

The most common method of elution development is under isocratic or isothermal conditions. This means that the composition of the mobile phase is constant together with the column temperature. Under these conditions, the interactive forces between any solute molecule and the two phases also remain the same and consequently, both the selectivity and the ultimate retention are also constant. However, for some mixtures which have a wide range of solute types, the elution of the most retained components can extend the analysis to an inconveniently long period of time. Consequently, there have been different methods of development introduced to excelerate the separation, in particular, the elution of the later peaks without impairing the resolution that has been obtained. They are as follows:

1. Flow Programming

This procedure can be used in either gas chromatography or liquid chromatography, but due to the sensitivity of many detectors to changes in mobile phase flow-rate, this procedure is not frequently employed. In this process the flow rate is increased continuously during the development of the chromatogram and thus excelerates the elution of the later and more retained solute peaks. Another disadvantage of this procedure is that the velocity of any solute band is only linearly related to the flow-rate, and to elute very strongly retained solutes would require inordinately high flow rates and consequently, inordinately high inlet pressures.

2. Temperature Programming

Temperature programming is used almost exclusively in gas chromatography and is a very effective method of excelerating the movement of solute bands along the column as the retention volume or retention time of a peak is proportional to the exponent of the reciprocal of the absolute temperature. Because the retention time is reduced exponentially, this development technique is extremely popular and a very common procedure in quantitative gas chromatographic analysis.

3. Gradient Elution

In liquid chromatography temperature programming is not as effective as in GC since the interactions in both the mobile phase and stationary phase are changed in the same proportion. Consequently, the retention of a solute is not nearly so sensitive to the operating temperature. The alternative, used in liquid chromatography, is gradient elution where the composition of the mobile phase is continuously changed during the development of the chromatogram. For example in the use of a methanol/water mixture with a reverse phase column in LC, the methanol concentration is usually continuously increased and as a result, the forces between the solute molecules and the mobile phase also increase and the strongly retained solutes are eluted more rapidly.

In gas chromatography, employing the flame ionization detector at a constant *elevated* temperature thermostated in a separate oven allows temperature programming to be carried out without changing the response of the detector. This means that temperature programming can be used satisfactorily for accurate quantitative analyses.

In liquid chromatography, only certain detectors can tolerate gradient elution (i.e. changes in mobile phase composition during development) and some of these will only function with certain solvents. The UV detector can be used with the gradient elution technique employing such materials as water, methanol, and acetonitrile as solvents, as they are all transparent at the UV wavelengths normally used for detection purposes. Consequently, the UV detector can be employed to provide accurate quantitative results with any gradient elution system using these solvents. The fluorescence detector is even more compatible with gradient elution providing none of the solvents themselves fluoresce at the excitation wavelength of the source and the solvents are free from fluorescing impurities. Unfortunately, the fluorescence detector has a very restricted linear dynamic range. In contrast, it is impossible to employ the refractometer detector for gradient elution due to the fact that the refractive index of the mobile phase will continuously change during the elution process and consequently result in extreme and unacceptable baseline drift.

THE FUNCTION OF THE CHROMATOGRAPHIC COLUMN–THIN LAYER PLATE

In the chromatographic column (or for that matter the surface of the thin-layer plate) two processes progress independently and concurrently. Firstly, as a result of the distribution of each solute between each phase, as already discussed, the individual solutes are separated during their passage through the column. At the same time, however, the local concentration of solutes will tend to disperse, and the column must be designed in such a manner that the solute band dispersion is contained sufficiently so that each solute peak is eluted discretely. A measure of peak dispersion is given by the column efficiency. The higher the column efficiency, the more narrow the peak and the better the column performance or resolution will be.

The column efficiency can be calculated for a given solute peak from the chromatogram. It is not germane to this book to describe the development of the equation that permits the efficiency to be calculated, but it is useful to give the basic procedure. The efficiency, in number of theoretical plates, is given by the following equation:

$$n = 4(y/x)^2$$

Referring to Figure 4, y is the retention distance between the injection point and the peak maximum and x is the width of the peak measured at 0.607 of the peak height. Peak width should always be measured using a comparator and the distance taken from the inside edge of the trace on one side of the peak to the outside edge of the trace on the other side of the peak. This procedure can help prevent errors resulting from the finite width of the trace.

Dispersion in a Packed Column

The efficiency of a column will vary with the column length and therefore, the goodness of the column has to be defined as some property/unit length so that valid comparison can be made between

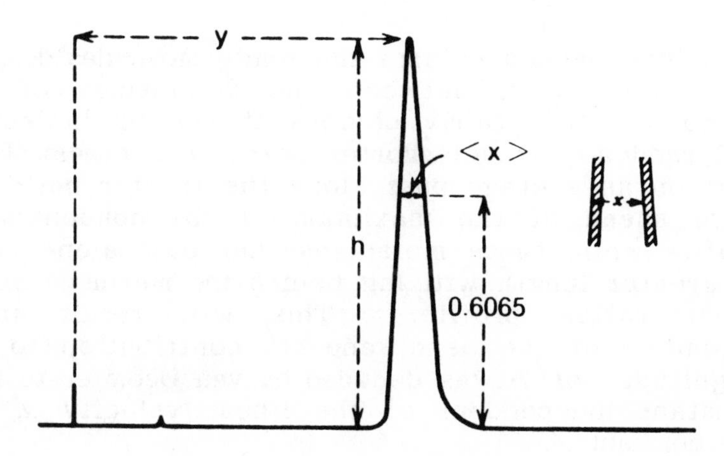

Figure 4. The Measurement of Efficiency.

columns. The property employed is the variance/unit length of the column *H* which can be calculated from the ratio of the column length *l* to the number of theoretical plates *n* of the column, i.e.

$$H = l/n$$

Again, it is not appropriate in this chapter to deal extensively with the equations that define *H* (variance/unit length) but some discussion of the factors that contribute to the magnitude of *H* would be pertinent. The best equation for a packed column (6) that describes *H* in terms of the velocity of mobile phase passing through it is that derived by van Deemter et al. (7) and takes the following form:

$$H = A + B/u + Cu$$

where u is the linear velocity of the mobile phase and A,B and C are constants.

Each expression in the equation pertains to a

specific dispersion process. The three processes are as follows:

1. The Multipath Process

In a packed column the solute molecules describe a tortuous path between the interstices of the support. It is fairly obvious that some molecules will randomly travel shorter paths than others. Those that on an average pass along the shorter path will move ahead of the maximum of the concentration profile while those molecules that pass along paths of greater length will lag behind the maximum of the concentration profile. This will result in a spreading of the band, and its contribution to the magnitude of H was deduced by van Deemter to be a constant independent of the linear velocity u, i.e. the constant A.

2. Longitudinal Diffusion Process

If a local concentration of solute is placed at the midpoint of a long tube filled with either a liquid or a gas, the solute will slowly diffuse to either end of the tube. It will first produce a Gaussian distribution with the maximum concentration at the center and finally, when the solute or solute vapor reaches the end of the tube, end effects will occur and the solute will diffuse until there is a constant concentration of solute throughout the whole tube. The latter effect is never realized in chromatography, but the initial spreading process does occur in the mobile phase in a column. The degree of spreading will obviously be proportional to some function of the time that the solute exists in the mobile phase and thus, if the mobile phase is flowing through the column at a linear velocity of u, the extent of the spreading process will be some function of $1/u$. The variance due to this spreading will also be directly dependent upon the diffusivity of the solute in the mobile phase and the geometry of the space occupied by the mobile phase. Van Deemter et al. showed that the variance due to this process in columns used for gas and liquid chromatography was inversely proportional to *u*, i.e. B/u.

3. Dispersion due to the Resistance to Mass Transfer in the Two Phases

During the movement of a solute band along a column, the solute molecules are continually transferring from the mobile phase into the stationary phase and back from the stationary phase into the mobile phase. This transfer process is not instantaneous because a finite time is required for the molecules to traverse the mobile phase in order to enter the stationary phase and *vice versa*. Thus, those molecules close to the stationary phase will enter it almost immediately, whereas those molecules some distance away from the stationary phase will find their way to it a significant time interval later. However, as the mobile phase is traveling at a given velocity along the column, the solute molecules in the mobile phase will move a finite distance along the column during this time interval. Thus, they will be absorbed into the stationary phase further along the column than those that were originally in close proximity to the stationary phase. The result of this delayed transfer of some solute molecules in the mobile phase leads to band spreading and is termed the band dispersion due to the resistance to mass transfer in the mobile phase. This dispersion effect is further amplified by the parabolic velocity profile of the mobile phase flowing between the particles that results from normal viscous flow. Thus, the mobile phase layer close to the stationary phase surface is static, and solute molecules that diffuse from this static layer into the moving fluid pass along the column at a rate dependent on the distance they diffuse into the bulk of the moving phase. Because of the parabolic nature of the velocity profile of the mobile phase between the particles, those molecules diffusing the greatest distance into the moving fluid move away from those molecules that diffuse only a small distance from the stationary phase surface thus causing band spreading or dispersion. There is a resistance to mass transfer in both phases and according to van Deemter, both are linearly related to *u*, i.e. Cu.

In GC (and under some circumstances LC) capillary columns are employed that contain the stationary phase as a thin film on the inside surface of the tube. The equation for *H* for a capillary column does not contain a multipath term *A* as there are no particles in the column and thus, the equation, as shown by Golay (8), takes the form:

$$H = B/u + Cu$$

An example of an experimental curve *H* versus *u* curve for a packed LC column is shown in Figure 5.

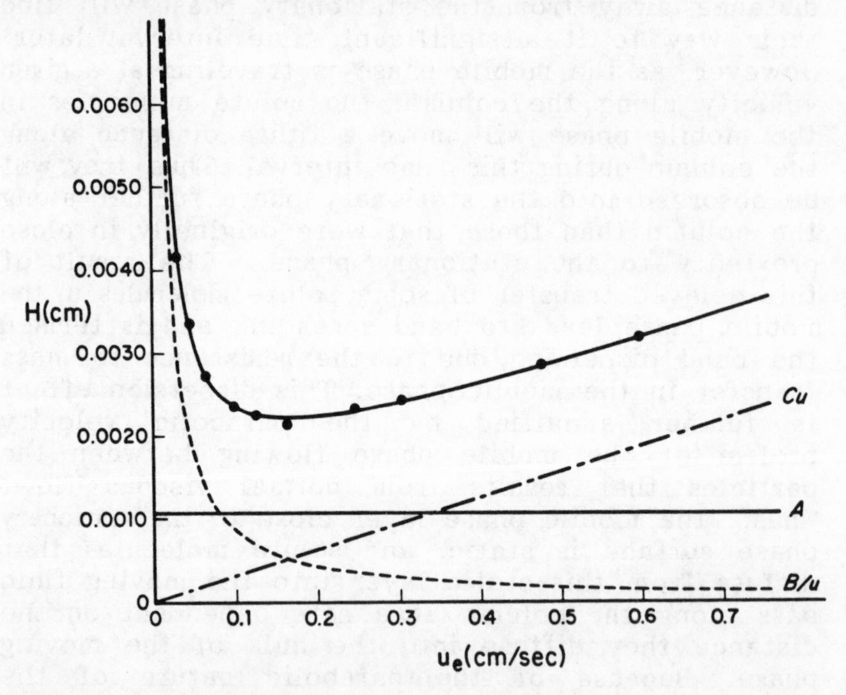

Figure 5. *H* versus *u* curve. Partisil-10; 5.4% ethyl acetate in n-hexane; benzyl acetate. Fit to van Deemter eqn.; r = 0.999699.

The experimental data shown was fitted to the van

Deemter curve and the magnitude of the individual terms calculated and also included in the diagram. The same type of dispersion takes place on the thin-layer plate and in principle should be described by a two-dimensional analog to the van Deemter equation. Unfortunately, however, the experimental equation for dispersion on the thin-layer plate that has been experimentally validated has not yet been developed.

Extra Column Dispersion

The total band width of an eluted solute is not solely dependent on the dispersion process that takes place in the column itself. Dispersion also occurs in the injection system, connecting tubes between injection valve and column, column and detector and in the detector itself. It is obvious that this extra column dispersion will effect the column efficiency and needs to be minimized. If not, the narrow solute bands that are eluted discretely from the column will merge together in the connecting tubes or in the detector cell itself.

Klinkenberg (9) suggested that extra column dispersion should be limited to 5% of that of the column dispersion or the band variance should not be increased by more than 10%. The band width of a solute eluted from a column will depend, among other things, on the column radius and thus, for a given detector system it will be necessary to determine the optimum column radius that will limit the increase in dispersion to 5%. This is equivalent to the reduction in column efficiency being restricted to less than 10%. In fact in practice, the radius of the column is increased to a level where the band width and hence the volume of mobile phase of a peak eluted from the column is 20 times the band width due to extra column dispersion. An explicit equation has been developed that can be used to calculate the optimum column radius for a given liquid chromatographic system and is given as follows (10):

$$r_c = \left(\frac{\sigma_e \sqrt{n}}{0.32\pi l \varepsilon} \right)^{\frac{1}{2}}$$

where r_c is the optimum column radius
σ_e is the extra column dispersion
n is the column efficiency
l is the length of the column
ε is the fraction of the column volume occupied by the mobile phase.

A similar equation for the radius of a GC column could be developed and would take a similar form. It is important that the radius of the column is not greater than that given by the above equation because the smaller the radius of the column, the greater the mass sensitivity (11). Consequently, the minimum column radius commensurate with the conditions suggested by Klinkenberg (9) should be the optimum to be strived for. The effect of extra column dispersion on mass sensitivity will be discussed later.

THE BASIC CHROMATOGRAPHIC SYSTEM

The Gas Chromatograph

A block diagram of the apparatus used in a basic gas chromatograph is shown in Figure 6. It consists of a pneumatic controller that will provide carrier gas for the column (normally helium in the USA, nitrogen in Europe) and any gas supply necessary for the detector, for example, air and hydrogen for the flame ionization detector. The next essential part of the chromatograph is the injection system which can be a septum device or a sample valve. Either injection device may be manually or automatically operated but needs to be thermostatted at the column temperature. The column is usually placed inside an oven which should have associated temperature programming facilities, provide a range of program rates and a maximum operating temperature of about 350°C. One type of detector is usually all that is necessary, but under special circumstances others may be required. The system must also include the necessary detector electronics that can provide an electrical output proportional to the concentration of solute in the gas eluted from the column. Most

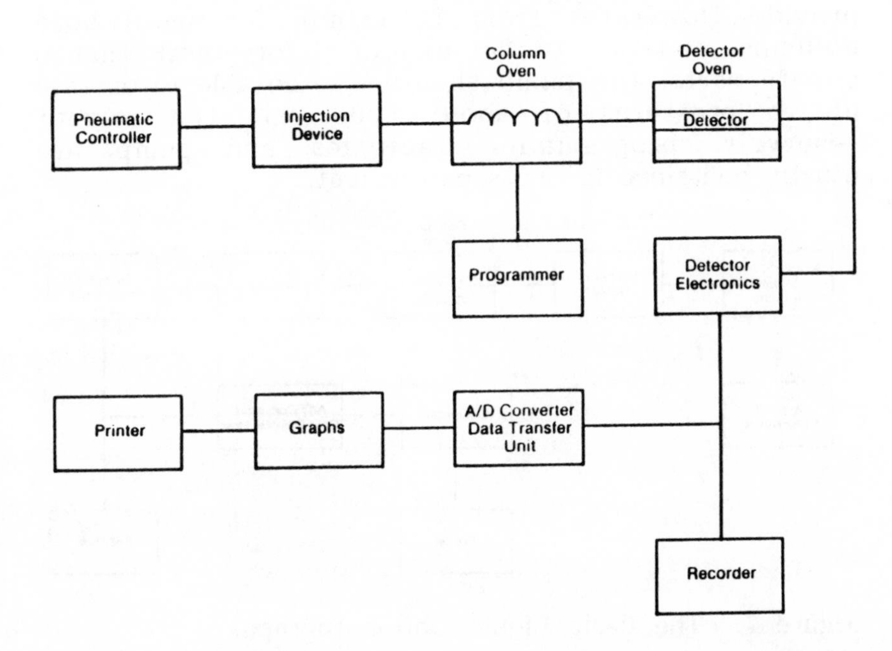

Figure 6. The Basic Gas Chromatograph.

modern chromatographs also have a data processing system that consists of an A/D convertor, a data transfer unit, a computer, and means of data printout. The analog output from the detector is usually passed to an appropriate potentiometric recorder. The system discussed is basic for quantitative analyses by GC but can be made far more sophisticated with multi-column and multi-detection facilities, disc memory for the computer and in some instances, automatic sample preparation devices.

The Liquid Chromatograph

A block diagram of the layout of a liquid chromatograph is shown in Figure 7. It consists of a mobile phase supply unit which can handle up to 4 different solvents but essentially requiring a minimum of 2. The system should include a suitable programmer, normally involving 2 solvents, but 3 or 4 are occasionally required. The pump employed should provide flow-rates from 10 μl/min for small bore columns and up to 5 ml/min for their macro counterparts, the pump should also be able to operate up to pressures of about 6000 psi. The solvent reservoir, programming facilities and pump are usually contained in one separate unit.

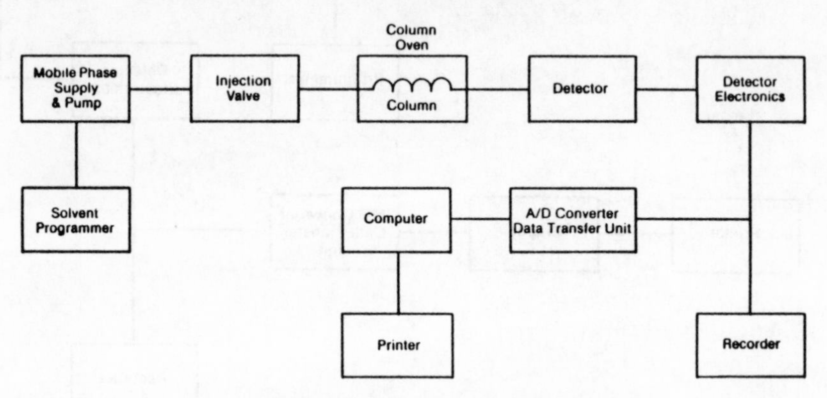

Figure 7. The Basic Liquid Chromatograph.

The pump feeds the mobile phase supply through a sampling valve which can have sample volumes ranging from .1 μl to 100 μl or even more and is sometimes fitted with an optional automatic sampling system. The output of the sample valve passes to the column which in general should be thermostatted, preferably with a high heat content fluid. In some cases, however, the column is held at ambient temperatures with no thermostatting device at all. The output of the column then passes to the detector and its associated electronics. In a similar manner to the gas chromatographic detector, the output from the amplifier is made proportional to the concentration of the solute in a mobile phase eluted from the column and is fed to a recorder and a parallel output to an A/D convertor, data transmission unit, and thence to a computer. The computer will usually be associated with an appropriate printout unit.

The liquid chromatograph described is basic for quantitative analysis by LC, but can also take a far more sophisticated form. As well as having automatic sampling facilities, a number of different detectors can be available or the detector can be very sophisticated and take the form of a variable wavelength detector or a diode array detector; multifunctional detectors are also available. The computer can also be supported by an appropriate disc memory.

Thin-Layer Plate System

The thin-layer plate system is simple in the extreme and consists merely of the plate and a trough for the mobile phase situated in an appropriate enclosure. Some of the advantages of thin-layer chromatography (TLC), (in fact, probably the basic reason for the survival of TLC as an analytical technique) result from its simplicity, the inexpensive equipment needed and its ease of operation. The most expensive part of the system is a scanner used for determining the density of the 'spots' to obtain quantitative results. Details of the use of such instrumentation for quantitative analysis will be given in a later chapter in the book.

THE ROLE OF THE DETECTOR

There are three basic properties of the detecting device used in any chromatographic system that has direct influence on the accuracy, precision and repeatability of any quantitative analysis. These detector properties are linearity, sensitivity and detector dispersion. A detail discussion on detectors will be given in later chapters but the basic requirements for accurate quantitative analysis will be briefly discussed here.

1. **Linearity**

Most manufacturers of detectors claim a linear response over a particular concentration range for any specific detector, but, in fact, linearity is a theoretical concept and can only be approached, and rarely, if ever, realized. Because of this, Fowliss and Scott (12) proposed a means of measuring linearity which can be defined by the following equation

$$y = ac^r$$

where y is the output from the detector
a is the constant
c is the solute concentration
and r is the response index.

For a truly linear detector, the response index of the detector r should equal unity and the proximity of r to unity gives a measure of the linearity. In the author's experience there has been no detector manufactured so far that has a value of r equal to unity over more than two orders of concentration range. In practice, the value of r should lie between .98 and 1.02 over three orders of magnitude of concentration range, if the detector is to be useful for quantitative analysis. However, from the point of view of quantitative accuracy, r does not have to be equal or close to unity *providing an accurate value of r is known*. If accurate results are important, it is strongly recommended that the manufacturer specifications are ignored and the true linearity is experimentally determined (13). Detectors used in GC, for example the FID, have linear dynamic ranges of perhaps five or more orders of magnitude. LC detectors rarely have linear range (as previously defined) over a range much greater than 3 or at most 4 orders of magnitude.

2. **Sensitivity**

There are three definitions of sensitivity which

are pertinent to a chromatographic system. They are the concentration sensitivity of the detector itself, the concentration sensitivity of the chromatographic system as a whole and the mass sensitivity of the chromatographic system. Discussions of the relative importance of these three different parameters are given in the book "Liquid Chromatography Detectors" by the author (11). It is sufficient to state here that the detector sensitivity is that concentration of solute in the mobile phase that gives a signal-to-noise ratio of 2. The chromatographic system, however, includes a column and other equipment components and thus, system sensitivity is more important than detector sensitivity, although the two are interrelated. The system concentration sensitivity is defined as that concentration which gives a peak having a height equivalent to twice that of the noise when the maximum volume of charge is placed on the column. (That is the maximum sample volume that still limits the extra column dispersion to not more than 10% of the column dispersion). In a similar manner, the mass sensitivity of the chromatographic system is defined as that mass of solute which, when contained in the maximum permissible sample volume, gives a peak having a height equivalent to twice the noise. From the point of view of quantitative analyses, the last definition of sensitivity (i.e.,the mass sensitivity of the chromatographic system) is the most important. It is given for a concentration sensitive LC detector by the expression

$$m = 2\sigma_c C_D$$

where m is the minimum mass detectable
σ_c is the dispersion due to the column
and C_D is the concentration sensitivity of the detector as previously defined.

The value of *m* has also been shown (10) to be given by

$$m = 6.25\sigma_e (1+k')C_D$$

where σ_e is the extra column dispersion

and k' is the capacity ratio of the solute.

The value of *m* is extremely important in those analyses where the quantity of the sample is severely limited.

A similar equation can be developed for a GC system using the FID, but the fact that the FID is a *mass sensitive* detector must be taken into account.

3. Detector Dispersion

In order to accurately determine all the individual components of the mixture they must be completely separated from one another in the column. As already discussed, to maintain the separation, the peaks must not be further dispersed in the connecting tubes or in the detector cell subsequent to leaving the column. It is therefore important that the detector dispersion (including connecting tubes and detector cell) is kept to a minimum. This allows the column resolution to be fully realized and the maximum amount sample volume to be injected onto the column (14). It has been already shown that the detector dispersion also controls the radius of the column. Consequently, as the radius of the column determines the column dispersion (σ_c), the detector dispersion also determines the mass sensitivity. Therefore, as it is important to have the most sensitive chromatographic system possible, the detector must be designed not only to be sensitive and have a linear response over a wide concentration range, but also to contribute minimum extra column dispersion to the eluted peak.

SYNPOSIS

Chromatographic analysis is popular because it is a procedure that *simultaneously separates the sample* into its components and *quantitatively determines the amount of each component* present. The *technique* has been *known* for over 80 years, but it was not until the *development of gas chromatography* in the early 1950's that it became an *established analytical technique.* The development of *liquid chromatography* rapidly followed in the early 60's and, together with thin layer chromatography, provided *unique analytical capabilities.* The chromatographic separation takes place between a *moving* and *stationary phase.* The different solutes are *retained* to *different extents* in the stationary phase by *exploiting different molecular interactions.* The molecular interactions can be divided into three simple groups, *ionic interactions, polar interactions* and *dispersive interactions.* There are three methods of chromatographic development, *frontal analysis, elution development* and *displacement development.* The vast *majority* of quantitative analyses in chromatography involves *elution development.* Elution development under *isocratic* and *isothermal* development results in *later peaks* taking an *inordinately long time to elute.* Consequently, techniques such as *flow programming, temperature programming* and *gradient elution* have been employed to *reduce the analysis time.* In *gas chromatography,* the most common is *temperature programming* and in *liquid chromatography, gradient elution.* The *goodness of a* column can be obtained by measuring its *efficiency* and to compare columns of different lengths, the *variance per unit length* is used which is taken as the *ratio* of the *column length* to *column efficiency.* There are three factors that *cause dispersion* in a column and thus denigrate the separation that is obtained, and they are the *multipath effect, longitudinal diffusion* and *resistance to mass transfer.* The relationship between variance per unit length and mobile phase velocity is best described by the *van Deemter equation.* The *dispersion* that can take place in *parts of a chromatographic system* other than the column is called *extra column dispersion. Extra column dispersion* should be minimized at all times as it *denigrates the separation* that has been achieved in the column. *Extra column dispersion* also determines the *minimum column radius* that can be used and this controls other properties of the chromatographic system. The *basic gas chromatograph* consists of a *pneumatic gas supply, injection system, column, detector* and *appropriate output monitoring device.* A *liquid chromatograph* consists of a *mobile phase supply, injection system, column, detector* and also *appropriate output monitoring system.* Having achieved the separation, the *detector* is the next most *important factor* in quantitative analysis. The detector *must be linear* and preferably, the linearity should be *quantitatively defined* in terms of the *response index.* Two of the most *important factors* that involve *chromatographic performance* are the *concentration and mass sensitivity.* The *dispersion* that takes place in the *connecting tubes* and *sensing cell* should be also *kept to a minimum.*

References

1. M.S. Tswett, Tr. Estestroispyt, Kazan Univ., No. 3 (1901) 35.

2. R. Kuhn and E. Lederer, Ber. Deut. Chem. Ges., 64 (1931) 1346.

3. J. Sawlewicz, T. Reichstein, Helv. Chim. Acta, 20 (1937) 949.

4. A.J.P. Martin and R.L.M. Synge, Biochem. J., 35 (1941) 91.

5. A.T. James and A.J.P. Martin, Biochem. J., 50 (1952) 679.

6. E. Katz, K.L. Ogan and R.P.W. Scott, J. Chromatogr., 270 (1983) 51.

7. J.J. van Deemter, F.J. Zuiderweg and A. Klinkenberg, Chem. Eng. Sci., 5 (1956) 271.

8. M.J.E. Golay, in "Gas Chromatography 1958",D.H. Desty (Ed.), Butterworths, London, 1958, p. 36.

9. A. Klinkenberg, in "Gas Chromatography 1960", R.P.W. Scott (Ed.), Butterworths, London, 1960, p.194.

10. "Small Bore Liquid Chromatography Columns", R.P.W. Scott, (Ed.) Wiley, New York, 1984.

11. R.P.W. Scott, "Liquid Chromatography Detectors", Elsevier, Amsterdam, 1986, p. 28.

12. I.A. Fowlis and R.P.W. Scott, J. Chromatogr., 11 (1963) 1.

13. R.P.W. Scott, "Liquid Chromatography Detectors", Elsevier, Amsterdam, 1986, p. 15.

14. R.P.W. Scott, in "Small Bore Liquid Chromatography Columns", R.P.W. Scott (Ed.), Wiley, New York, 1984, p. 9.

Quantitative Analysis using
Chromatographic Techniques
Edited by Elena Katz

Chapter 2

DETECTION IN QUANTITATIVE LIQUID CHROMATOGRAPHY

Kenneth Ogan

The quantitative utility of chromatography stems not only from the spatial segregation of the various components of an unknown sample, but also from the generation of an output signal that is *quantitatively* related to the amount of each sample component present. Hence, the analytical value of the chromatographic system depends in no small measure on the performance characteristics of the detecting system. The successful utilization of an analytical chromatographic system requires that the specifications of the detector be appropriate for the analyte levels expected.

DETECTOR PERFORMANCE

Linearity

The detector generates an output signal which increases or decreases in response to increases or decreases in the solute concentration in the eluent. There is a *minimum* solute concentration below which the detector signal cannot be discerned due to noise, and an *upper* solute concentration beyond which the detector response is saturated (i.e. a maximum electronic output signal). The concentration range encompassed by these two limits is the usable range for the system, across which there is a predictable change in the detector output in response to a change in solute concentration injected. Use of solute concentrations outside this range will obviously result in loss of quantitative information. An output signal that changes linearly with solute concentration changes greatly simplifies the quantitative interpretation of the chromatogram (and indeed, this is the only relationship which is suitable for data interpretation by manual means); the range of concentration values across which this linear relationship holds is defined as the *linear dynamic range* of the system. The exact definition of what constitutes a linear response varies. Some authors have proposed that a detector be defined as linear if the predicted and observed signal levels differ by no more than 5%. This criterion is strongly dependent on the specific data set used to extrapolate the detector response. Scott, in his classic book on LC detectors (1), defines a detector *response index*, as the exponent which results in the best fit of the detector response to varying solute concentration,

$$y = ac^r \tag{1}$$

where y is the detector signal,
c is the solute concentration,
a is a proportionality constant,
and r is the response index.

This response index can be easily determined by plotting log(y) vs. log(c); the slope of this plot is just the exponent *r*. Scott makes the recommendation that $0.98 \leq r \leq 1.02$ for a "linear" detector.

Sensitivity

The lower limit of the analytical range, i.e. the lowest concentration that can be detected, is set by the observed noise in the system. This noise may originate in the detector itself, or it may originate from other sources in the chromatographic system. This lowest detectable concentration is a very important characteristic of the system since it is a measure of the *ultimate sensitivity* of the analytical system. However, it is also one of the most misused, and misunderstood, of chromatographic characteristics. There are several different descriptive parameters which relate to the lower detection limit; different authors have developed different descriptions of these parameters, and have sometimes even given a different name to a parameter previously named otherwise. We shall return to the discussion of the minimum concentration that can be detected after some additional nomenclature is introduced.

A recent A.C.S. committee report (2) made recommendations on definitions and nomenclature that are becoming widely accepted. First, the lowest concentration of an analyte that can be detected with an analytical method is termed the *limit of detection (LOD)*. The LOD is referenced to the total analytical method, and therefore will be affected by factors such as sample losses in sample preparation steps. However, this term can equally well be used in reference to just the analytical chromatography step where appropriate. (Other terms used for this parameter include minimum detectable level, detection level, minimum detectable amount.) This A.C.S. committee report also defines a *limit of quantitation (LOQ)*, which is the concentration level above which concentrations can be determined unambiguously. This report recommends that the LOD be determined as the signal corresponding to 3 times the peak-to-peak noise (or 3 times the standard deviation of the noise distribution if proper blanks are available), while many other authors have used the signal that is 2 times the peak-to-peak noise for the limit of detection. Although the detector generates a signal proportional to the concentration of solute in the eluting peak, chromatographic factors such as column efficiency and capacity factor govern the volume of the peak, with the result that the output signal is influenced by both detector and chromatographic performances. Peak heights (and peak areas) change in response to changes in the injected concentrations of the respective solutes. For a chromatographic system being operated in the linear mode, the *volume* of any given peak is constant, independent of the concentrations of the injected solutes, and the change in output signal reflects the different solute concentrations in the solute peak (or the change in the rate of solute mass flow, for those rare LC detectors that are mass sensitive rather than concentration sensitive; see Chapter 1).

The detector contribution to the total chromatographic performance can be extracted by assuming a general peak shape, as described by Scott (1) in his book on LC detectors. The concentration at the maximum of a Gaussian peak (or a near-Gaussian peak, or even a triangular peak) is approximately twice that of the average solute concentration in the peak. The average solute concentration in the peak is the total mass injected (m) divided by the peak volume. The peak volume for a Gaussian (or near-Gaussian) peak can be taken as 4σ where σ is the standard deviation for the peak. (Mathematically, the σ for a peak is determined from its second moment. However, this complex computation can be avoided if the peak shape is Gaussian or near-Gaussian, in which case σ can be determined from the peak width. The peak width can be measured at any point along the peak, but is most conveniently taken at 0.607 of peak height where the width is 2σ.) The *detector response* (R_C) is defined as the increase in detector output for a unit increase in solute concentration in the detector cell (for a linear detector; see Chapter 1), and is given by

$$R_C = h/(2m/4\sigma) = hw/m \qquad (2)$$

where h is the peak height,
m is the total mass injected (the product of the solute concentration in the sample and the injection volume), and
w is the peak width at 0.607 peak height *in volume units*

(i.e. peak width (in cm), divided by chart speed (in cm/min), multiplied by flow rate (in ml/min)).

This formulation corrects for chromatographic effects on the solute concentration in the detector cell. A column having low efficiency will result in a much lower peak height, but will have a correspondingly much wider peak width. (Detector response, defined here according to Scott (1), is equivalent to the term, "sensitivity" used by some other authors.)

In principle, once the detector response has been determined, the detector sensitivity, or *minimum detectable concentration* (C_D) can also then be determined from a measurement of the detector noise (n_D)

$$C_D = \beta\ n_D/R_c \quad (3)$$

where β is a scalar, typically taken as 2, but sometimes as 3 by some authors, as noted earlier. The actual measurement of the noise level is not as simple as it might appear. First, there is the question of exactly what to use for n_D; i.e. whether it should be a simple peak-to-peak measurement, or the standard deviation of the noise, or some other noise amplitude value. Second, there is the question of how long baseline ought to be sampled in order to obtain a trace that is characteristic of the true noise. Scott (1) recommends use of the maximum peak-to-peak fluctuation observed over a period of 10 minutes, while the A.S.T.M. E19 committee recommends a 15-minute period (3). Knoll (4) has examined this question from the point of view of the information capacity of the chromatogram. He derived relationships specifying the appropriate length of baseline monitoring required, expressed in terms of multiple peak widths, together with the corresponding noise amplitude factor.

TABLE 1

Multiplier Constants for Various Baseline Sampling Times

Baseline Period	Ω* $\beta = 2$	$\beta = 3$	Time**
10 $w_{\frac{1}{2}}$	0.77	0.99	2.1 min
20 $w_{\frac{1}{2}}$	0.57	0.72	4.1
50 $w_{\frac{1}{2}}$	0.37	0.46	10.3
100 $w_{\frac{1}{2}}$	0.26	0.33	20.5

*Multiplier to be used in relating noise amplitude to minimum detectable concentration during the measurement period (see Eq. (3a)).

**Assuming column dimensions 25 cm long, 0.46 cm diameter; 1 ml/min flowrate, 10,000 plates (10 μm particles), k'=2.

His results relating the appropriate scaling factor for various measurement periods are summarized (and extended) in Table 1. (Knoll elected to use the peak width at half-height, $w_{1/2}$, rather than the width at 0.607 peak height. The width at 0.607 peak height is 0.85 $w_{1/2}$.) The noise is expressed as the largest peak-to-peak fluctuation (n_D) within the observation period, and Table 1 gives the multiplier to be used with this noise amplitude to generate C_D according to

$$C_D = \Omega n_D / R_c \tag{3a}$$

with Ω taken from Table 1. Table 1 also includes the baseline observation time necessary for the case of a column 25 cm long and 0.46 cm in diameter, packed with 10 μm diameter particles (with an assumed efficiency of 10,000 plates), with a capacity factor of 2 for the peak of interest, and using a flow rate of 1.0 ml/min. These latter conditions are those given by Scott (1) in conjunction with his recommendation of monitoring the baseline. It is seen that the multiplier to be used in equation 3a is strongly dependent on the time taken to monitor the baseline. It should also be noted that columns of different dimensions generate peaks of different volumes; hence, the characteristic time across which the drift and noise are to be monitored must be appropriately adjusted.

System Noise

This determination of the detector sensitivity involves the measurement of the system noise level and as such will include noise generated by other parts of the chromatographic system in addition to that of the detector. The lower limit of detection will certainly be set by the system noise level, but this is not necessarily the lowest limit of which the detector itself is capable. The true detector sensitivity will be fixed by the inherent noise of the detector, but this is a very difficult parameter to determine. One might be tempted to turn off all other parts of the chromatographic system in order to obtain the "true" detector noise, but this will inevitably alter the thermal state of the detector, and the thermal responsiveness of the detector is one source of system noise. Effects of pressure fluctuations would also be lost with this approach. At present, the only viable solution for the analyst is to note carefully the chromatographic conditions under which the noise measurement is made, and further, to request from the manufacturer the numerical values for the detector response to thermal and pressure changes (i.e. the magnitude of the change in detector signal in response to a unit change in pressure or temperature). Thermal effects can be troublesome, especially for high-speed chromatography, because of the heat generation from viscous dissipation (5,6). Inadequate pulse dampening can also be a troublesome and annoying noise source. Yet another source of extraneous system noise is impurities in the mobile phase. In the following, we make the assumption that the system is functioning properly, and that all possible precautions have been exercised; however, the noise remains potentially a system noise and not purely detector noise.

DATA HANDLING

The utility of chromatography comes from its separation of the components of a sample mixture. Quantitative success depends on the accurate and precise measurement of peak area or peak height. The solute concentration is determined by comparison with similar measurements in chromatograms from injections of the pure solute at a variety of concentrations. Implicit in this procedure is the assumption that the specific peak in the analytical chromatogram is due solely to the solute of interest. Peak identity cannot be unambiguously established by use of retention time (volume) alone. The peak capacity of any chromatographic system (*i.e.* the number of fully resolved peaks that will fill the chromatogram) is finite and limited, and the odds are that the chromatogram of any general sample mixture will not even begin to approach the regular spacing involved in the description of peak capacity (7). Consequently, independent means must be utilized to further characterize a chromatographic peak for quantitative purposes. Good quantitative results are obtained by careful work and attention to detail on two fronts; the accurate and precise determination of peak area or peak height, and unambiguous peak identification and confirmation of peak purity. The choice of the detector and its proper use play an important role in both of these areas as it will be discussed in Chapter 3.

Peak Heights and Peak Areas

Quantitative analysis utilizes the peak height or the peak area to determine the concentration of a specific solute in the sample. Quantitation requires that the *response* (R_c) of the detector to the solute be fully established, *i.e.* that the calibration curve has already been determined. (A linear response is virtually always assumed, although non-linear responses could in principle be accommodated with the use of a properly programmed computer.) Manual quantitation measurements are inevitably based on peak height measurements, but today, most all quantitative determinations utilize a computer, either to measure peak height or peak area. Peak area is obtained by summing the *net* detector signal from the start of the peak to its end. In practice, this summation procedure is carried out on a time base, rather than with respect to the elution volume which is the true dependent chromatographic variable. The detector signal is sampled at frequent, regular time intervals (set as some integer multiple of the computer clock period), and the average (or maximum, or final) detector signal in this period, corrected for the baseline value, is then added to a running sum. Each individual signal is related to the total solute mass in that peak segment (the time increment is equivalent to volume increment; concentration multiplied by volume is the amount of solute),

$$A_p = \Sigma c_i \Delta t = \Sigma(m_i/\Delta v)\Delta v = \Sigma m_i = m \qquad (4)$$

where A_p is the peak area,

Δv is the volume corresponding to the time slice, Δt,

c_i is the solute concentration in the time slice $(c_i = m_i/\Delta v)$

and m is the total mass of the solute injected.

Thus, peak area is theoretically independent of chromatographic parameters, and hence, independent of variations in column performance from run to run, and from day to day. (In fact, it is even independent of the type of column or the type of chromatography.)

Peak height, in contrast, is directly related to the column efficiency, and to the capacity factor. Changes in column performance with time will produce changes in peak height that are not related to the concentration of the solute in the sample. However, peak area is accrued on a time basis, while chromatographic development is based on elution volume. Measurements of peak area implicitly assume a perfectly steady flow of mobile phase. If the flow changes, or fluctuates, the detector will continue to generate a signal proportional to the concentration of solute in the cell in each time interval, independent of the increment in elution volume. (This is the case for concentration sensitive detectors, which include most all LC detectors.) A variable, or erratic, mobile phase flow rate will yield irregular peak areas. Peak height, on the other hand, is unaffected by changes in flow rate (for concentration sensitive detectors), and hence would be the preferred measurement for systems exhibiting irregular mobile phase flow (due to inadequate pulse dampening or marginal pump performance).

The use of peak area also assumes that the beginning and ending of the solute peak are accurately determined in all cases; hence errors are introduced by the inaccurate assignment. In practice, it is sometimes found that quantitation based on peak height is more precise for the earliest eluting peaks in a chromatogram, while quantitation based on peak area is always more reproducible for later peaks. In fact, this is likely due to the greater relative error introduced by inaccurate assignment of the starting and ending points for the narrow peaks at the beginning of a chromatogram. The accurate assignment of these two points is the key to successful quantitative liquid chromatography.

The procedures and variables available to the analyst using computer-based chromatographic peak interpretation are many and varied, depending on the specifics of the system being used. A few processes are common to most all systems, while further embellishments and options may, or may not, be available. Much of the flexibility of the specific system depends on the amount of memory available. Because of the variety of features available in these units, only the general characteristics will be described in this chapter. The interested reader is referred to more detailed treatises, and to the manufacturers' information sheets.

Signal Digitization

The analog detector signal must first be converted into a digital form that is then transmitted to the computer (or microprocessor). There are four or five different approaches to the actual digitization step (analog-to-digital, or A.D., conversion), but most systems utilize either a dual-slope A/D converter, or a voltage-to-frequency converter. (See ref. 8, for example, for more detailed information.) A parameter of major importance is the *resolution* of the A/D converter, which is expressed in "bits". A bit is the basic unit of binary information, typically expressed as either a "0" or a "1", *e.g.* the absence or presence of a reference voltage level. The output of an A/D converter is a string of bits, expressing the amplitude of the converted signal in binary format. The number of bits used to express this value affects the precision with which this amplitude is determined. Thus, a hypothetical 1-bit A/D converter would generate either a "0" or a "1", indicating only whether the signal amplitude were above or below the mid-range point. A 2-bit A/D converter would generate one of 4 ($=2^2$) values, "00", "01", "10", or

"11", thereby increasing resolution by a factor of 2 over the 1-bit converter. Similarly, a 4-bit A/D converter would represent a further improvement by a factor of 4, with 16 (=2^4) possible different values (and each value being expressed as a string of 4 bits). The number of bits in the A/D converter then represents the coarseness of the digital representation of the analog signal. In order to achieve a precision of better than 1%, the A/D converter must have at least 7 bits in resolution, while to achieve a precision of 0.1% (which is the minimum suitable for chromatography), the converter must have 10 bits in resolution (see Table 2). The higher the number of bits provided by the A/D converter, the more expensive and complicated it is.

An alternative to the direct, "brute force" A/D conversion utilizing a large number of bits is the use of an auto-ranging amplifier preceeding the A/D converter. Step changes in gain are automatically executed to keep the signal input to the converter within a limited range. The output string of bits then includes the digitized level as well as a coded value representing the appropriate gain value. This allows the use of a simpler A/D converter, but requires the complexity and expense of the auto-ranging amplifier.

TABLE 2

Potential Error Due to the Least Significant Bit

#Bits (N)	2^N	Least Sig. Bit as %
4	16	6.2%
6	64	1.6
8	256	0.4
10	1024	0.1
12	4096	0.02
14	16384	0.006

The technology in this area is continuing to change rapidly, and neither approach holds an unassailable position.

Table 2 lists the maximum precision to be expected for A/D conversions at different bit levels. These are *maximum* precision values, for full scale input signals. A peak with a maximum that is 1/10 full scale will have the same *absolute* uncertainty (the least significant bit), but the *relative* uncertainty will be much larger because the signal itself is so much smaller. Hence, the percentage values given in Table 2 are the best-case values, and the user ought to plan to add at least two more bits in order to obtain the necessary levels of precision.

A second important characteristic of the digitization step is the *frequency* at which the input signal is sampled. If the input signal is sampled at too few points across the peak, accurate recovery of the peak height or peak area is impossible. As a general rule of thumb, a *minimum* of 10 points across the peak should be acquired in order for the accurate measurement of peak parameters. The peak volume depends on the specific details of the column dimensions (Chapter 3); hence the sampling rate must be set for each column type. Peak width increases with elution volume in isocratic chromatographic development, and even with gradient development, no peaks will be narrower than the early eluting peaks; hence, the peak width of the first eluting peaks sets the sampling rate for the system. Some data systems will progressively slow the data acquisition rate as the run progresses, thereby saving data space; however, this mode of operation obviously must be suppressed for gradient operation in order to record accurately the narrow peaks occurring as the gradient develops.

One major difference between data handling systems is whether the chromatogram is processed in real time (or "on-the-fly") and only selected characteristics such as retention time and peak area stored, or whether the whole chromatogram is stored in digital format for later processing (making possible repeated processing as necessary). The first approach suffers from the inability to re-process the chromatogram with altered peak detection parameters, while the latter approach obviously requires considerably more memory than the first.

Sources of Error

Quantitative measurements require the accurate and precise determination of peak height and peak area.

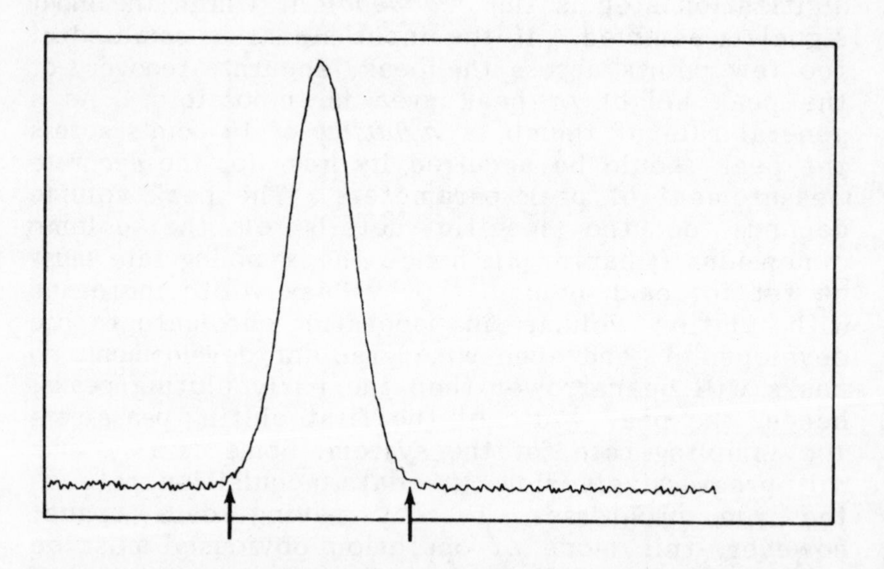

Figure 1. Simulated Chromatographic Peak. Gaussian Peak Shape S/N = 50; Integration Threshold = 2 x (Noise Level).

The measurement of peak area in particular is subject

to a number of difficulties and pitfalls and yet, as was indicated in the introduction, peak area is less subject to variations in column performance than is peak height. The determination of the *start* of the chromatographic peak, and of the *end* of the peak, is a primary source of error. The importance of these two parameters, peak start and peak end, can readily be demonstrated through the use of a computer generated peak with added random noise.

Gaussian Peaks

The peak start and the peak end points for a *Gaussian peak* indicated in Figure 1 are typical of those seen in many analytical publications.

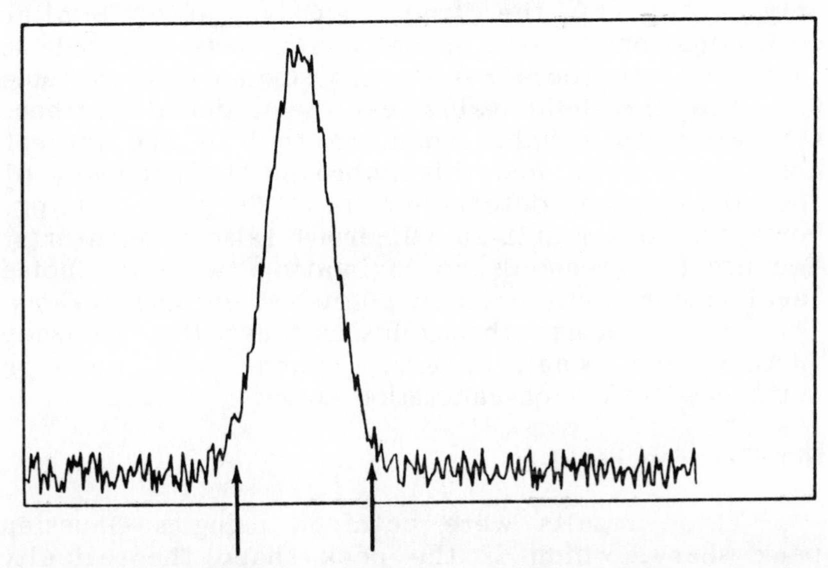

Figure 2. Simulated Chromatographic Peak. Gaussian Peak Shape S/N = 10; Integration Threshold = 2 x (Noise Level).

One approach to the determination of the peak start is to set a threshold level slightly above the maximum noise amplitude observed, such that when the detector output increases above this level, the data handling station software treats the signal as the beginning of a peak. Subsequent digital values are summed until the end of the peak is detected. In the simplest case, the end of peak signal occurs when the detector signal amplitude drops below the same threshold value used to establish peak start. This approach was used in Figure 1 with the threshold set to be 2 times the maximum noise amplitude. The effect of this threshold setting on a peak having an amplitude just at the limit of detection (LOD, S/N = 10) is shown in Figure 2. The peak start and peak end are unambiguously determined (which would not be the case if the threshold were set at precisely the noise amplitude). However, the peak area thus obtained from the settings in Figure 2 accounts for only 97% of the true Gaussian area. Similar determinations can be made as the peak amplitude is increased and these results are summarized in Table 3. The threshold value can be reduced further, ultimately to a value equal to that of the largest noise excursions, and this improves the accuracy of the peak area determination. Such a setting, however, does result in numerous false peak starts. The use of thresholds approximating twice the noise level is commonly seen in published chromatograms. The use of larger thresholds reduces the accuracy further for small peaks, which will have a detrimental effect on calibration curves.

Non-Gaussian Peaks

These results were obtained using a Gaussian peak shape, which is the peak shape theoretically expected, and observed, under ideal conditions. However, *non-Gaussian peaks* are frequently encountered, and integration errors are more severe for them. Tailing peaks can arise due to excessive extra-column volume, to overloading of the column, or to heterogeneous retention mechanisms. The theoretical peak shape resulting from excessive extra-column volume is an exponential function

TABLE 3

Ratio of Measured Peak Area to True Peak Area for Different Integration Threshold Settings

	Gaussian Peak			Exponentially-Convoluted Gaussian Peak		
Signal-to-Noise Ratio	Threshold* 1.5	2.0	3.0	Threshold* 1.5	2.0	3.0
3	0.908	0.900	0.821	0.853	0.774	0.674
5	0.959	0.953	0.898	0.893	0.858	0.835
10	0.990	0.972	0.951	0.942	0.933	0.903
15	0.995	0.987	0.981	0.958	0.954	0.932
25	0.995	0.993	0.985	0.983	0.983	0.960
50	0.997	0.995	0.993	0.993	0.994	0.991
100			0.997	0.995	0.994	0.991
200						0.995

*(in units of noise amplitude)

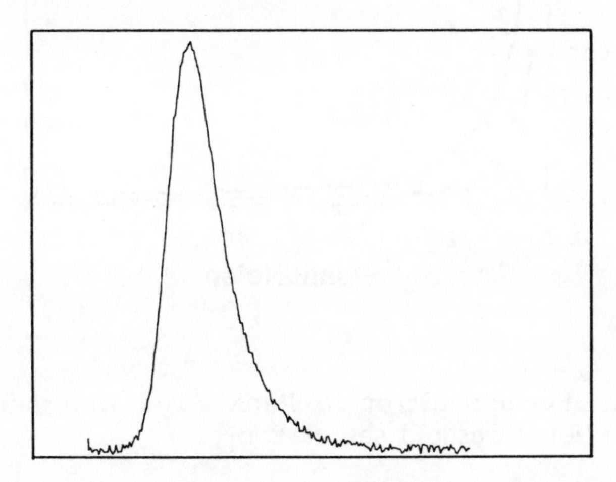

Figure 3. Gausssian peak convoluted by an exponential function.

convoluted with the Gaussian function, such as shown in Figure 3. This function is also a good empirical model for peaks distorted by other factors. The long, drawn-out tail of these peaks makes the accurate identification of the peak end more difficult and prone to error. Even with the threshold set to just above the noise level, a 6% error results at the LOD, in contrast to a 1% error for a pure Gaussian peak (see Table 3). Use of a more realistic 2x, or even 3x, the noise level as the peak threshold results in errors as large as 7% and 10%. These are shown graphically in Figure 4. The effect of these errors on the calibration curves is correspondingly larger as well.

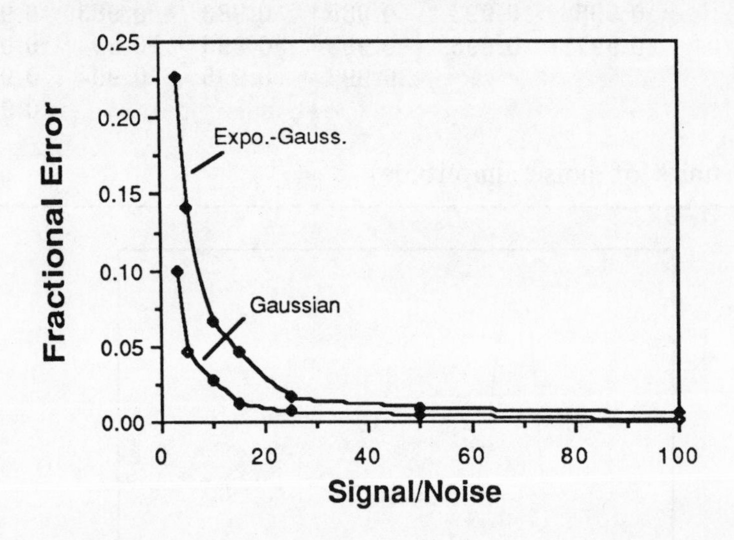

Figure 4. Percent Error in Peak Area Determination for integration threshold set at 2 n_D

These errors can be significant; for peaks ranging from S/N = 3 to S/N = 200, the apparent slope is increased by 7.5%, 5%, and 4% for threshold settings of 3x, 2x, and 1.5x the noise level, respectively. (The corresponding errors in the slopes for purely Gaussian peaks are 5%, 2.6%, and 2%, respectively.)

Baseline Drift

The use of a simple threshold value for peak start and peak end becomes impossible when the detector baseline is drifting or otherwise changing. A solution to this problem is to use the *first derivative* of the *detector signal* as a means of better determining the true peak start. In Figure 5, a Gaussian peak has been added to a sloping baseline; the corresponding first derivative is also shown. To the extent that the detector drift is constant, the baseline for the first derivative of the detector output will be constant, in principle making it easier to detect the peak start. However, Figure 5 is the ideal case in that there is no detector noise.

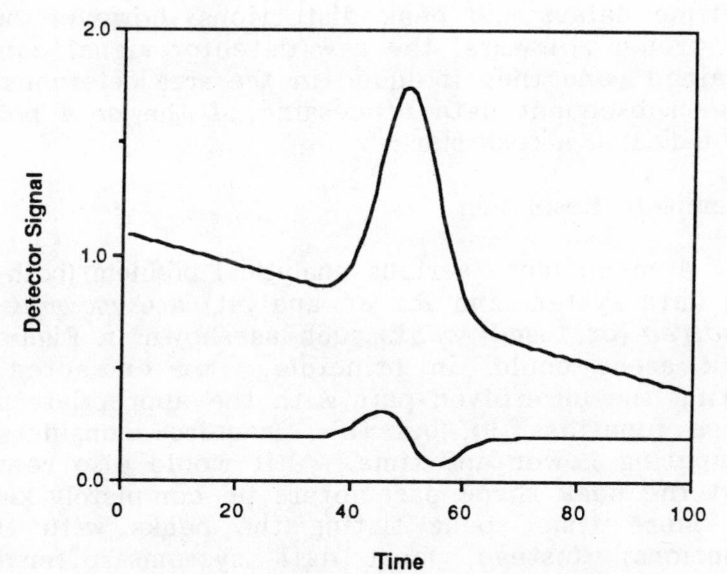

Figure 5. Drifting Gaussian Peak and Derivative

As those who have worked with derivative systems can well testify, the use of a derivative function adds considerable noise to the signal. For example, a S/N factor of 30 in the original peak would be reduced to approximately 5 in the derivative signal.

There are three main ways of reducing the observed noise. The first is to filter the detector signal in order to reduce the noise amplitude (however, there are limits to the extent of which the detector signal can be filtered while still retaining accurate peak shape). Another option is to use a running average, which can be easily implemented in software; sequential digitized detector signals are averaged together. Equal weighting need not be given to all values within the group, e.g. the Savitsky-Golay smoothing algorithm for which the coefficients are typically -3, 5, 12, 5, and -3. Finally, the data can be "bunched", in which successive *n* points are added together and treated as a single data point. All of these approaches result in time delays and peak distortions; however, with the proper software, the raw detector signal can be retained and then included in the area determination after subsequent data processing of the data points has indicated a peak start.

Incomplete Resolution

A much more serious analytical problem (both for the data system and for an analyst) are *incompletely resolved* (or fused) *peaks* such as shown in Figure 6. Peak areas could, in principle, be extracted by fitting the unresolved pair with the appropriate peak shape functions (9), but this requires considerable computing power and time. It would also require that the peak shape parameters be completely known (or more time spent fitting the peaks with trial functions). Instead, most data systems offer the option of either splitting the area with a vertical line dropped from the valley between the overlapping peaks to the baseline, or skimming the second peak, in which a projection is made from the valley between peaks to a tangential point on the tail of the second peak.

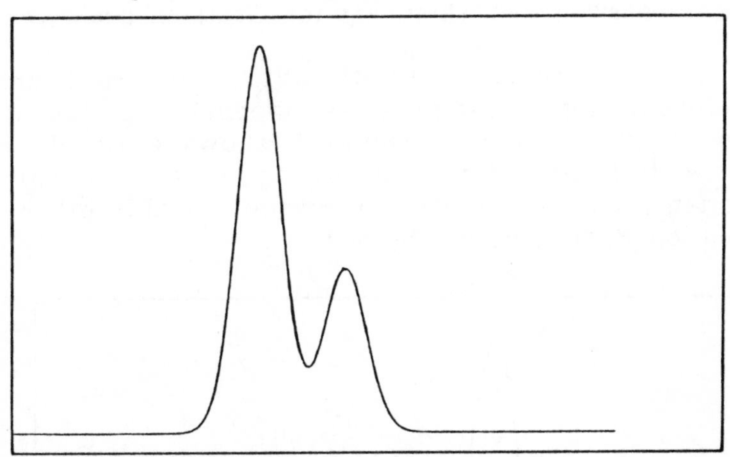

Figure 6. Fused Chromatographic Peaks. Gaussian peak shape, peak separation = 4σ.

Some data systems offer several different baseline treatments with respect to the first option.

The error inherent in these two approaches can be evaluated with computer generated peaks. The accuracy of the experimentally determined peak areas for fused peaks depends on the resolution between the two peaks, and on the relative heights of the two peaks. For example, for the pair of peaks shown in Figure 6, with R_S = 1.0, the error in the area of the second peak using the vertical drop method varies from -10% for a peak height ratio of 0.10 (the first ratio for which there is a distinct valley) to -0.3% for peaks of equal heights. The experimental error in the area for the first peak of the pair ranges from +1.4% to +0.3% for the same sets of peaks. In contrast, the errors in the areas for the second peak obtained by skimming the same fused peak pairs ranges from -64% for the 0.10 peak height ratio to -28% for peaks of equal height. The corresponding errors in the area determination for the first peak when the second peak area is obtained by skimming range from +6% to +29%. The vertical drop always gives a more accurate answer than skimming for Gaussian peaks.

It is more difficult to obtain accurate areas for fused, non-Gaussian peaks because of the long tails of these peaks. Figure 7 shows a fused peak pair with a Gaussian peak riding on the tail of an exponentially convoluted Gaussian peak, all with resolution R_s=1.0, as in Figure 6.

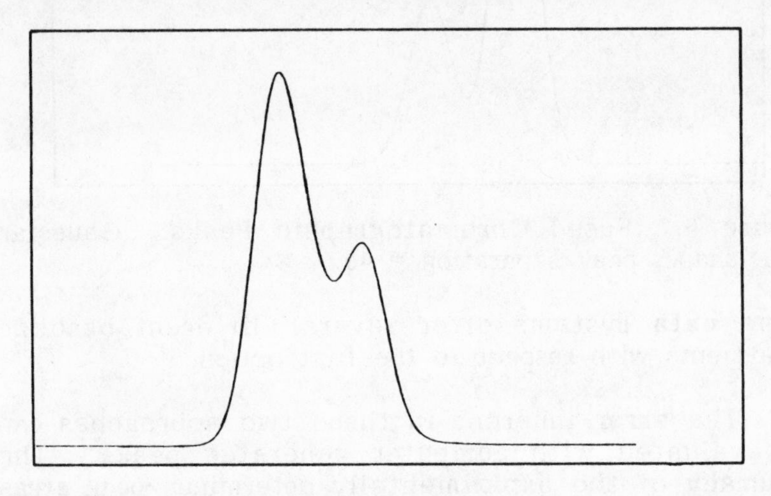

Figure 7. Fused Chromatographic Peaks. Exponential leading peak (τ = 1.5σ) with Gaussian peak following (peak separation = 4σ).

The experimental error for the area of the second peak of this fused pair ranges from +80% for a peak height ratio of 0.30 (again, the first peak height ratio for which there is a distinct valley), to +39% for equal peak heights. The corresponding errors in the determination of the area of the first peak range from +12% to -22%. The use of skimming for this fused pair results in errors of -71% to -35% for the second peak area, and +12% to +21% for the first peak area. The errors in the skimming procedure are thus smaller than those for the vertical drop, but they are nevertheless very large and significant for both methods.

The experimental errors in the determination of the area of fused non-Gaussian peaks are so large as to shed doubt on the usefulness of such methods for quantitative purposes. In Figure 8, the area of the second peak, as obtained by vertical drop and by skimming, is plotted against the true peak area.

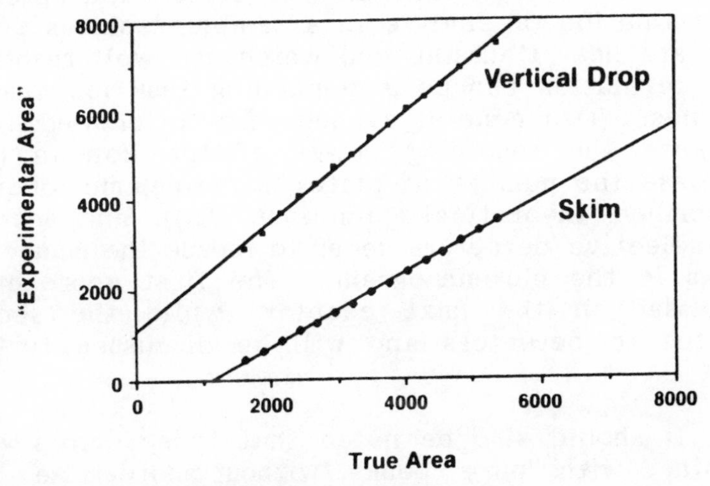

Figure 8. Fused Chromatographic Peaks.

The linear relationship between the "experimental" and the "true" areas for both methods means that the experimental area can be related to the amount of solute injected, provided that a calibration curve similar to that in Figure 8 is constructed with injected solute concentration as abscissa. Such plots must be done because of the non-zero intercepts; simple proportional calculations clearly would be erroneous. The accuracy of the experimental peak areas is not high, but because the error is also linearly related to the solute concentration, a correlation between peak area and the concentration of injected solute can be obtained. (The non-zero intercept for these fused non-Gaussian peaks also indicates that the method of "standard additions" would not be applicable in such cases.) Implicit in the development of the plot in Figure 8 is the assumption that the area of peak #1 remains constant. If peak 1 is also varying, then the correlation between peak area and solute concentration will break down, and large errors in the determination will result.

It is seen that the accurate and precise determination of analyte in a sample requires peaks that are near Gaussian and which are well resolved. High resolution can be a demanding task for complex samples. Two general avenues can be followed; first, increase the resolving power of the column (*i.e.* increase the number of plates in the column following prescribed theoretical guidelines (10)), and second, use selective detectors so as to reduce the number of peaks in the chromatogram. The first approach is discussed in the next chapter, while the second relates to detectors and will be discussed further here.

It should also be noted that these errors were obtained with "pure" peaks, without added noise. The errors described in the previous section on peak start and peak end in the presence of noise suggest that additional errors of several percent will be introduced for real peaks with attendant noise.

PEAK IDENTIFICATION

Multiple Detectors

The successful use of a data station or other means to measure peak area or peak height for quantitative results requires that the peak(s) being measured be due solely to the analytes of interest. The measurement of peak height or peak area is but one part of the total quantitative measurement; additional methods and techniques are required to ascertain the identification of the peaks. Several of methods are available to the analyst for the purpose of confirmation of peak identity, including repeated analysis on a column of a different type, alternative chemical treatment of the sample (to either increase or decrease the amount of analyte or to alter its chemical nature), and finally the use of multiple detection techniques in order to characterize further the solutes as they elute from the column. Because this last approach requires no further manipulations or handling of the sample, it tends to be the more useful and popular approach. The simplest execution is to merely add detectors in series at the end of the column. In actuality, there are only a limited number of different LC detectors and careful consideration must be given to their selection.

The major detectors in use today for LC are the refractive index (RI) detector, the UV absorption detector, the fluorescence detector, and the electrochemical detector. The RI detector monitors the change in angle due to refraction of a light beam as it passes from one medium to another. The extent to which a light beam is refracted is a function of the refractive indices for the two different media, and if one medium is the LC eluent stream, the appearance of solutes in this stream causes the refractive index to change and consequently to change the direction of the refracted light beam. (Note that this process reflects the bulk property of the eluting stream, rather the specific properties of the solute itself.) There are two different basic types of refractive index LC detectors (1). The RI detector approaches being a truly universal detector since the presence or absence of most any solute alters the refractive index of the liquid medium. The RI detector is consequently widely used and is particularly useful for the monitoring of solutes which do not absorb UV light, are not fluorescent, or are not electrochemically active. The RI detector suffers from a sensitivity to flow rate perturbations and to temperature fluctuations. Also because it is a bulk property detector, it is not as sensitive as other spectroscopic or electrochemical detectors.

The UV absorption detector is by far the most common detector in use in LC detection today. Light is directed through the sample stream eluting from the end of the column and the amount of the light absorbed by solutes in the eluent is monitored. This light can be either a fixed wavelength or a broad band of wavelengths, in which case a filter or a monochromator is used to select the monitoring wavelength. Not all analyte species absorb UV radiation and hence the UV detector is not nearly as universal as the RI detector. It does, however, offer good sensitivity, good stability, and a ruggedness in design and operation that has made it the premier LC detector today. Today, the user not only can choose between a fixed wavelength and a variable wavelength UV detector, but can also select a diode-array UV detector (see below).

The fluorescence detector and the electrochemical detector are not nearly so widely used as the UV absorption detector but are very useful detectors in those applications in which they can be used. The fluorescence detector monitors the light emitted by solute molecules in response to absorbance of exciting UV light. Because not all molecules fluoresce, the fluorescence detector is a much more selective detector than either the RI detector or UV detector. It offers very high sensitivity, low noise and it is fairly rugged in operation. The electrochemical detector is truly "orthogonal" to both the UV and RI detectors in the sense that its signal is not related to spectral properties, but rather reflects the oxidative or reductive tendency of eluting solute molecules. Most electrochemical detectors are operated as amperometric detectors in which the voltage is fixed between two electrodes immersed in the eluent stream and the current required to maintain this fixed voltage is monitored as a function of time. The presence of a solute which can be oxidized or reduced between these two electrodes (at the fixed potential) results in the generation or consumption of an electron at one electrode as the solute is either oxidized or reduced. Electrochemical detectors offer very high sensitivity, comparable to that of fluorescence detectors. Also, they are very selective since not all compounds are readily oxidized (or reduced).

The most useful application of multiple detectors is the coupling of selective detectors which only respond to a restricted set of the analytes in the sample. In this way, a much simpler chromatogram is generated from each of the selective detectors. A set of overlapping or "fused" peaks seen by a near-universal detector is then replaced by sets of individual peaks which, as noted in the previous section, provide much higher analytical accuracy and precision. Furthermore, the appearance of the peak in the selective detection helps to confirm the identify of the peak.

Multi-channel Detectors

A highly selective detector can be very helpful by generating well resolved chromatographic peaks, but it can also be problematic since not all compounds of analytical interest necessarily respond to a single setting of the selective parameter of the detector. For example, not all analytes might respond to a particular excitation and emission wavelength choice for a fluorescence detector. This would then require multiple chromatographic runs, or changes in detector conditions during the course of a single run, or other approaches in order to successfully monitor all of the analytes of interest. One alternative approach is to use more sophisticated selective detector which generate multiple channels of information. The diode-array UV detector is a prime example of such an approach (see for example, ref. 11). In the diode-array detector, the single detecting element of a UV detector has been replaced by an array of solid state detecting elements (photo-diodes). The optical system has been changed such that the dispersed spectrum is in focus across the array of diodes. Each diode element of the diode array then corresponds to a wavelength window. This enables a wide spectral segment to be continually monitored as the chromatogram develops. These detectors typically have a large number of diodes in the array (256, 512 or 1024 elements) which means a considerable amount of analytical information which must be digested by the data processing system. One approach to this mountain of data is to select only a few wavelengths and to monitor them continuously as the chromatogram develops, rather than trying to record several hundred wavelengths. The selectivity of the UV detector is not as high as that of a fluorescence or electrochemical detector. UV spectra consist of broad bands, and distinguishing between two similar spectra can be difficult. One useful means of comparison is to compute the ratio of the observed absorbances at two wavelengths within the band. This ratio is constant and distinct for different compounds. An excellent example of this approach was recently provided by Ramnaraine and Tuchman (12) who reported the absorbance ratioing for 57 different aromatic and nitrogen acids of interest in biochemical and clinical analyses.

Multi-channel detectors analogous to the diode-array UV detector have also been developed for fluorescence and electrochemical detectors. The electrode potential in the electrochemical detector can be rapidly swept and the electrochemical current recorded as a function of voltage (rapid-scan voltammetry, e.g., ref. 13), exactly analogous to the recording of absorbance as a function of wavelength in the diode-array UV detector. The excitation and emission wavelengths of a fluorescence detection can be scanned synchronously [with a constant energy difference] (14), which generates spectral information analogous to that from the diode-array detector. There is also the option of varying the excitation and emission wavelengths independently (15), which offers an additional degree of freedom relative to that of the diode-array and rapid-scan voltammetric detectors. Such detectors provide a wealth of information for the analyst, but do require extra data acquisition and processing capability. The rapidly falling price of memory and computational power means that significant developments in peak identification techniques can be expected in the next few years.

SYNOPSIS

A detector must be *operated* within its *dynamic range*, that is between that concentration at which the signal is discernible from the noise and that concentration at which the detector fails to give a change in response. For *accurate quantitative analysis* the detector must be operated within its *linear dynamic range,* that is where the relationship between detector output and solute concentration is closely linear. The limit of detection, LOD, of the system is that concentration or mass that provides a peak having a defined signal-to-noise ratio the magnitude of which can be unambiguously defined. The *response* of the detector is the *voltage output per unit change in concentration* and the sensitivity or minimum detectable concentration is taken as proportional to the ratio of the detector noise to the detector response. The proportionality constant is commonly taken as two or three. *System noise* includes *short-term* noise, *long-term noise* and *drift.* Long-term noise is the most serious and is often caused by ambient changes and changes in flow rate.

Peak identification cannot be assumed from retention data alone and accurate analysis requires positive peak identity. *Quantitative analysis* requires the *measurement of peak heights or peak areas.* Peak height measurements depend on the constancy of chromatographic conditions but not strongly on constant flow rate; peak area measurements are nearly independent of chromatographic conditions but are very dependent on flow rate. Accurate peak area measurements also rely on the accurate identification of the start and end of a peak. Peak start and peak end can be identified from an absolute change in detector signal or by a change in its derivative. *Chromatographic data requires appropriate analog-to-digital conversion* and the quantitative resolution that is obtained depends on the number of bits associated with the conversion. Another important aspect of the data processing is the rate of data acquisition which will depend on the elution rate of the chromatographic peaks. In general, *a data acquisition rate of 10 data points per second is the minimum* necessary for accurate analysis. The most common *sources of error* are caused by the *inaccurate identification* of the *peak start* and *peak end* and the area assessment of *unresolved peaks.* Assignment of peak areas by vertical drop from the valley between peaks and by peak skimming can both provide serious errors depending on the peak shape and extent of resolution. Adequate resolution is the only assurance of minimum error. Peak *identification* is best *achieved by multiple detection* preferably by simultaneous detection processes. This can be attained by the use, for example, of the diode-array detector or by coincident or sequential multifunctional detection. Specific detection such as fluorescence detection can also simplify the separation and reduce the need for high resolution.

References

1. R.P.W. Scott, "Liquid Chromatography Detectors",

Elsevier, Amsterdam, 1986.

2. A.C.S. Committee on Environmental Improvement, Anal. Chem., 52 (1980) 2242.

3. ASTM Committee E19, No. E685-79.

4. J.E. Knoll, J. Chromatogr. Sci., 23 (1985) 422.

5. H. Poppe, J.C. Kraak, J.F.K. Huber and J.H.M. van den Berg, Chromatographia, 14 (1981) 515.

6. E. Katz, K. Ogan and R.P.W. Scott, J. Chromatogr., 260 (1983) 277.

7. J.M. Davis and J. C. Giddings, Anal. Chem., 55 (1983) 418.

8. C.E. Reese, J. Chromatogr. Sci., 18 (1980) 201.

9. S.D. Frans, M.L. McConnell, J.M. Harris, Anal. Chem., 57 (1985) 1552.

10. E. Katz, K. Ogan and R.P.W. Scott, J. Chromatogr., 289 (1984) 65.

11. A.F. Fell, H.P. Scott, R. Gill and A. C. Moffat, J. Chromatogr., 273 (1983) 3.

12. M.L.R. Ramnaraine and M. Tuchman, J. Chromatogr. Sci., 24 (1985) 549.

13. J.J. Scanlon, P.A. Flaquer, G.W. Robinson, G.E. O'Brien, and P.E. Sturrock, Anal. Chim. Acta, 158 (1984) 169.

14. M.J. Kerkhoff and J.D. Winefordner, Anal. Chim. Acta, 175 (1985) 257.

15. I.M. Warner, M.P. Fogarty, and D.C. Shelly, Anal. Chim. Acta, 109 (1979) 361.

Quantitative Analysis using
Chromatographic Techniques
Edited by Elena Katz

Chapter 3

QUANTITATIVE ANALYSIS BY LIQUID CHROMATOGRAPHY

Raymond P. W. Scott

There is a dichotomy of purpose in any chromatographic analysis as already stated in Chapter 1. The chromatographic column first separates the individual components of the mixture into a series of discrete peaks. Then, by the use of an appropriate detector and data evaluation procedure, the mass of solute associated with each peak can be identified. It cannot be stressed too strongly that the success and accuracy of the quantitative evaluation of the mixture depends heavily upon the success of the initial separation procedure. There are a number of very sophisticated algorithms that can be employed for computer processing chromatographic data that purport to provide accurate peak area measurements for individual solute peaks that are only partially resolved. Indeed, under some circumstances, such methods may be necessary where complete resolution of the components of the mixture,for some reason or other, is too difficult to achieve. This should be, however, the exception rather than the rule and the emphasis should be placed on achieving satisfactory resolution rather than resorting to mathematical manipulation to compensate for inadequate chromatographic performance. In general sophisticated mathematical algorithms are a poor substitute for good physical chemistry.

Ideally each solute peak should be completely resolved from its neighbor and exhibit perfect Gaussian form. Employing contemporary columns, correct analytical procedures and good chromatographic hygiene, ideal conditions can be closely approached for most liquid chromatography

(LC) analyses. For maximum accuracy and precision, the distance between the peak maxima of the closest eluted pair (the critical pair) should be as great or greater than 5σ (five standard deviations) which will result in almost baseline separation; the minimum resolution recommended is a peak separation of 3σ. Even with the most sophisticated deconvolution algorithms, such a resolution can lead to serious errors, particularly for solute pairs that have significantly different peak heights. In Figure 1 pairs of solute peaks represented are separated by 2σ, 3σ, 4σ, 5σ and 6σ, respectively.

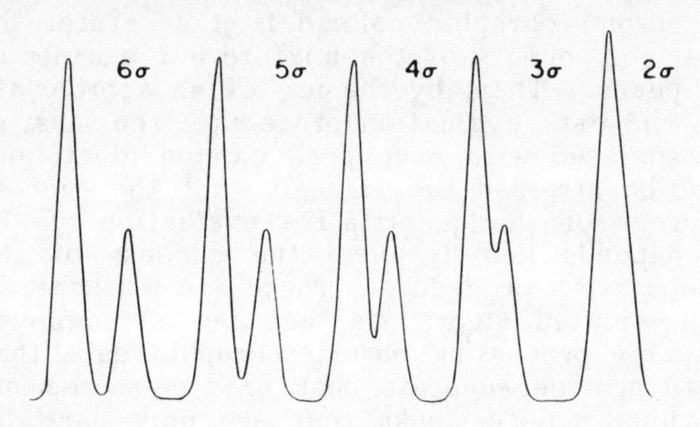

Figure 1. Peaks Demonstrating Different Resolution.

In order to rationalize the treatment of the subject of quantitative LC, this chapter will be divided into six sections. It will commence with general suggestions on the operation of a chromatograph for optimum quantitative accuracy and precision followed by a discussion on detector selection and the choice of detector operating parameters. Methods of sample preparation will then be described followed by injection procedures. Finally the basic procedures for quantitative analyses by LC will be outlined and levels put forward that indicate the precision that may be expected from LC analyses.

OPERATING PROCEDURES FOR QUANTITATIVE LIQUID CHROMATOGRAPHY

Should the LC system be in modular form, then the components of the chromatograph, pump, injection system, column, detector, recorder and data processing system should be arranged on the bench in an orderly and convenient manner. All controls should be readily accessible, in particular the sample valve should be in a position to be easily and rapidly operated if a manual version is being used. To minimize dispersion, all connecting tubes should be kept as short as possible and if necessary coiled. The area around the chromatograph should be kept clean and free from old sample vials, derivatization equipment and other paraphernalia that tends to accumulate round any chromatograph that is being actively used. This is just one aspect of chromatographic hygiene that is so necessary to ensure accurate and precise results and the importance of chromatographic hygiene will be continually referred to throughout this chapter.

The ambient temperature should be kept between 18°C(64°F) and 30°C(86°F), not merely for personal comfort, but because this is the operating range of most chromatographs. There are instruments that operate at temperatures as high as 45 °C(113°F) but these are rare and are usually designed for tropical use. It will be seen later that for accurate retention measurements the temperature of the column must be carefully controlled and a column thermostat is strongly recommended preferably with a high heat content thermostating medium (1). Provision should also be available for preheating the mobile phase prior to entering the sample valve and column. It should be noted that the sample valve should also be situated in the oven.

There is, for any LC analysis, an *optimum column*

that will separate the sample in the minimum time, with the minimum solvent consumption and the maximum mass sensitivity (2). For a specific analysis carried out on a given chromatographic system, the optimum column will have a unique combination of column length, column radius and be packed with particles of a particular diameter. Furthermore, the column will be operated at a specific optimum flow rate and have a predictable maximum sample volume. Thus if an optimized column is used for the analysis, the best operating conditions will have already been defined. However, in many instances the optimum column will not have been constructed and it is more likely that a commercial column is chosen which is close to the optimum. In fact, it may well be that the column used will be the result of arbitrary choice or even chosen solely on the basis of availability. It follows that some advice on general operating conditions for unoptimized columns might be useful. The column should be operated at flow rates close to that which give the optimum linear velocity according to the Van Deemter equation (3,4,5). Unfortunately, this optimum velocity will vary with the nature of the sample, stationary phase, the capacity ratio of the first solute of the critical pair (5) and the column radius. However, approximations can be made to provide some guidance as to the selection of the flow rates for a particular column geometry.

The simplified Van Deemter equation takes the form of:

$$H = A + B/u + Cu \tag{1}$$

where A, B and C are constants,
H is the variance per unit column length
and u is the linear mobile phase velocity

Differentiating equation (1) and equating to zero, the optimum value of u that provides the minimum value of *H* (maximum efficiency) is seen to be given by the following equation:

$$u_{(opt)} = (B/C)^{1/2} \quad (2)$$

where $B = 1.2\ D_m$
$C = (1+6k'+11k'^2)\ d_p^2/(24(1+k')^2 D_m)$
and D_m is the diffusivity of the solute in the mobile phase which normally takes a value lying between 1.5×10^{-5} and 3.5×10^{-5} cm^2/sec giving an average of about 2.5×10^{-5} cm^2/sec
k' is the capacity ratio of the first eluted solute of the critical pair
and d_p is the particle diameter

TABLE 1

Optimum Linear Velocity for Columns Packed with Particles of Different Sizes and Solutes Eluted at Different k' Values

	Linear Velocity cm/sec			
d_p (μm) / k'	3	5	10	20
2	0.205	0.125	0.062	0.031
4	0.182	0.109	0.055	0.027
6	0.174	0.104	0.052	0.026
8	0.169	0.101	0.051	0.025
Mean	0.183	0.110	0.055	0.027

Generally k' ranges from about 2 to 8 and LC packings are presently available in 3, 5, 10 and 20 micron sizes. Consequently, employing equation (2) the range of optimum values of *u* can be calculated and the results are shown in Table 1.

It is seen from Table 1 that although the

optimum linear velocity varies significantly with particle diameter, the effect of the value of k' is minimal and an average value can be taken for each particle diameter without serious error. From the average value for the optimum velocity for each particle size, the flow rate can be calculated by the following equation:

$$Q = 60\pi r^2 \varepsilon u \qquad (3)$$

where Q is the flow rate in ml/min
r is the column radius
and ε is the fraction of the cross-sectional area of the column available for flow and is usually taken as 0.75.

Employing equation (3) the recommended flow rates can be calculated for columns of different radii, packed with particles of different diameter and the results obtained are shown in Table 2.

TABLE 2

Optimum Flow Rate for Columns of Different Radii Packed with Particles of Different Diameters

	Flow Rate ml/min					
r(cm)	0.05	0.10	0.15	0.20	0.25	0.30
d_p (μm)						
3	0.065	0.259	0.582	1.03	1.62	2.33
5	0.039	0.156	0.350	0.622	0.972	1.40
10	0.019	0.078	0.174	0.310	0.484	0.697
20	0.010	0.039	0.087	0.135	0.242	0.349

It is seen that very small flow rates may be necessary if small bore columns are employed, particularly if packed with particles of relatively large diameter. It is also seen that the flow rate range of the pump must at least encompass flow rates from 10 µl/min to 2.5 ml/min if adequate chromatographic versatility is to be realized.

A major source of error in both peak height, peak area and even retention time measurements arise from *detector noise* resulting from flow rate fluctuations. It follows that it is important to employ either a pulse-free pump or insert an appropriate pulse dampener between the pump and the injection valve. An example of the effect of noise on the precision of retention time measurements was furnished by Scott and Reese (6). These authors chromatographed the same simple mixture twelve times to determine the precision of a computer data acquisition system for measuring retention times. The apices of the two peaks having the extreme values for retention time are shown in Figure 2. The importance of noise on the measurement is clearly seen. The time difference between the extreme values is seen to be 4.4 sec.

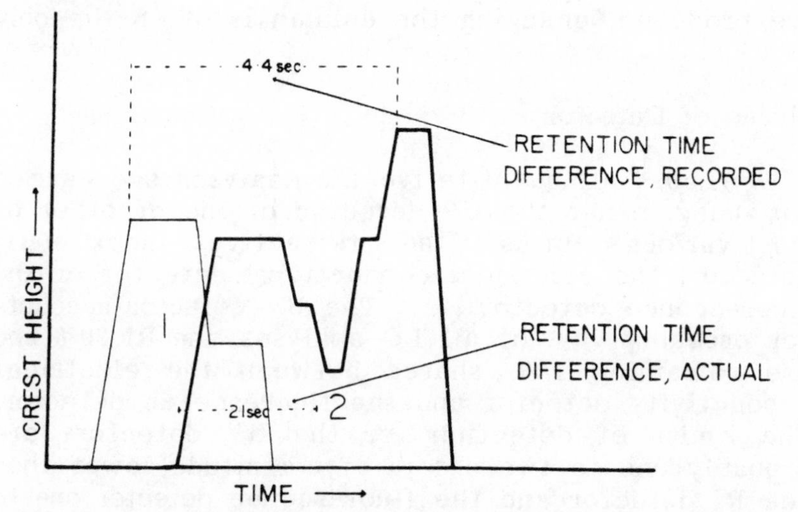

Figure 2. Peak Crests Reconstructed by the Computer (99.99-100% Peak Height).

This results from the fact that a noise spike

occurred at the front of the peak with the minimum retention value and a similar noise spike occurred on the rear of the peak with the maximum retention value. Consequently, the true maximum of the peak was not sensed. In fact, as seen in the diagram, the true time difference between the peaks was only 2 sec. Variation in flow rate is not the only source of noise from a chromatographic system; column contamination from sample residues after extensive use can cause serious noise. Incomplete equilibrium with a new mobile phase is another common source of noise, particularly drift. After changing the mobile phase, adequate time for equilibrium should always be allowed before commencing analysis. To prevent contamination by sample residues, the column should be regularly cleaned with appropriate solvents. Contaminated columns can also be a source of peak asymmetry, but more often it is due to inhomogeneous site activity or voids in the stationary phase. Cleaning the column with solvents can help, however, if the peaks begin to tail seriously, or double peaks are produced, changing the column is often the only solution.

Choice of Detector

Almost all quantitative LC analyses are carried out using either the UV detector in one or other of its various forms, the refractive index (RI) detector, the electrical conductivity detector or the fluorescence detector (7). The UV detector accounts for probably 70% of all LC analyses, the RI 20% and the remaining 10% shared between the electrical conductivity detector and the fluorescence detector. The limits of detection for the UV detectors are probably one to two orders of magnitude lower than the RI detector and the fluorescence detector one to two orders of magnitude lower than the UV detector. The electrical conductivity detector has similar detection limits to the UV detector. The UV, RI and electrical conductivity detectors have linear dynamic ranges of at least three orders of magnitude whereas the fluorescence detector may have a linear range of little more than two orders of magnitude. In practice the restricted linear range of the fluorescence detector is often tolerated for the rate of its high sensitivity.

The *UV detector* is probably the detector of choice for quantitative analysis as it combines the essential features of wide linear dynamic range with fairly high sensitivity. The solutes of interest must, of course, absorb in the wavelength range emitted by the UV source. It is possible, by means of derivatization techniques (8), to render a solute 'UV visible' that does not absorb in the UV. The derivatization process can easily introduce quantitative errors and consequently these techniques should be avoided wherever possible. Another advantage of the UV detector is that it can be used with gradient elution in reverse phase systems as the solvents normally employed are UV transparent. In contrast the use of the UV detector in forward phase separations places serious restrictions on the choice of mobile phase. Rabel (9) recommended a series of solvents that could be used with the UV detector in forward phase chromatography but even these, although helpful, give significant baseline drift during gradient development.

The *refractive index detector* is probably the closest to the universal LC detector available today. It does, however, due to it being a bulk property detector, have a limited sensitivity and consequently is less popular than the UV detector. It cannot be used with gradient elution and is very sensitive to changes in flow rate or mobile phase temperature. It is, nevertheless, very useful for detecting those solutes that do not possess a UV chromophore and in particular in polymer analysis. Providing the polymer contains more than ten monomer units, the refractive index of the solution is directly proportional to the concentration of the polymer. Consequently, quantitative analysis can be carried out by normalizing the peak areas.

The remaining commonly used detectors are selected for specific solute types. It is obvious that the *electrical conductivity detector* would be selected for the detection of ions and the *fluorescence detector* for those compounds that naturally fluoresce. However, due to the high sensitivity available from fluorescence detectors, derivatization procedures are often carried out to render other solute types fluorescent and thus detectable at low concentrations. Consequently, the relatively restricted linear dynamic range of the fluorescence detector is accepted and even the errors introduced by the derivatization procedure tolerated in order to utilize its advantageous low detection level.

The most important specification of an LC detector is its *linearity*. All commercial detectors are claimed to be 'linear' but, in fact, linearity is a theoretical concept and the response of practical devices only *tend* to be linear. The analyst requires to know how close to true linearity a detector is and the range over which the 'linearity' extends. Fowlis and Scott (10) suggested a rational method for defining linearity by the use of the response index (r) which has been discussed in previous chapters.

The value of r is, in fact, a measure of the linearity. If the detector was indeed truly linear, then r would be equal to unity and the proximity of r to unity indicates how closely the response of the detector tends to linearity. Unfortunately detector manufacturers do not provide values for r and so, at this time, the analysts are left to determine it themselves. This measurement, however, is not difficult and simple methods have been described (7). In practice, the value of r should lie between 0.98

and 1.02 if reasonable accuracy is to be obtained by assuming the response to be linear. In Figure 3 curves are shown relating detector output to solute concentration for different values of r. It is seen that individually, all the curves appear to be straight but the errors that can arise from this assumption, are shown in Table 3. It is seen that for a binary mixture and for values of r of 0.97 and 1.03 errors of nearly 6% can arise in the component present at the 10% level.

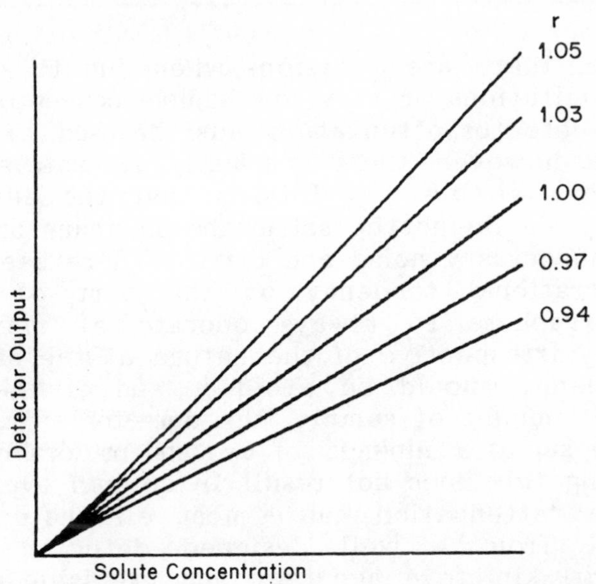

Figure 3. Graphs of Detector Output against Solute Concentration for Detectors Having Different Response Indices.

It follows that for very accurate work the value of r must be known and if the value differs significantly from unity, appropriate corrections must be made.

Perhaps surprisingly, very high sensitivities are not frequently required in quantitative LC.

TABLE 3

The Analysis Of A Two Component Mixture Using Detectors Having Different Response Indices

Solute	r = 0.94	r = 0.97	r = 1	r = 1.03	r = 1.05
1	11.25%	10.60%	10.00%	9.42%	9.05%
2	88.75%	89.40%	90.00%	90.58%	90.95%

Of course, there are occasions when due to sample volume limitations or very low sample concentrations, minimum detector attenuation must be used. For the most part, however, there is usually adequate amount of sample available. It follows that the detector sensitivity should not be set at the maximum and this avoids unnecessary noise and drift. There seems to be an irrational tendency on the part of many chromatographers to always operate at maximum sensitivity irrespective of the nature of the sample. This tendency should be avoided, and if there is sufficient amount of sample, the detector attenuation should be set at a minimum of two and preferably 4 or 8 providing this does not result in column overload. Significant attenuation can almost eliminate noise and drift from a well designed detector and consequently improve accuracy and precision. If a computer data acquisition system is being employed and the detector attenuation is not controlled by the computer, then the threshold level of acquisition must be increased in line with the attenuation. The most accurate and precise results will be obtained in the absence of all noise and drift (6, 11, 12).

Sample Preparation

All samples presented for LC analysis will normally require some preparation which may range in complexity from simple dilution with an appropriate solvent to an involved derivatization procedure. In many instances liquid samples may need one or more treatments such as filtration, solvent extraction, concentration or derivatization before it is ready for analysis. Solid samples can require even more involved preparation procedures that can be both lengthy and tedious, for example, tablet analysis. As the technique of LC becomes more widely used, many of the more involved sample preparation procedures are likely to be taken over by the laboratory robot. The robot has already become firmly established in many pharmaceutical control laboratories for LC sample preparation providing faster analyses, higher precision and greater accuracy. No doubt as LC techniques become more extensively used in the food and biotechnology industries, they too will make use of the robot for sample preparation.

The various procedures used in sample preparation are unique to the particular sample and so specific preparation routines will not be discussed here. Instead, some general comments will be given and some recommendations made.

Liquid samples often contain solid material which can block sample tubes, sample valves, columns and detector connecting tubes. To ensure chromatographic hygiene, the sample should be filtered, if not 'as received', then the sample solution should be filtered prior to injection. Many samples contain traces of strongly retained substances that can accumulate on the column and cause deterioration. To avoid this situation and maintain a clean column, a guard column can be used, situated between the sample valve and the column. The guard column is usually a small bore column packed with a pellicular material coated with the same stationary phase as that used in the column. Strongly retained materials are absorbed by the guard column and prevented from entering the analytical column whereas the solute of interest passes through it and onto the column. The guard column should be replaced regularly otherwise contaminants will eventually elute through it and defeat its purpose.

Solid samples will always require dissolution and almost invariably some of the solid will not dissolve, consequently, a *filtration* step is essential. Despite this preliminary filtration, if there are subsequent preparation stages to be carried out prior to the preparation of the final solution, this should again be filtered before injection. These filtering stages may extend the preparation time a little but this time will be more than compensated for by the reduced 'down-time' of the chromatograph.

Some liquid samples may require *concentration* before they are suitable for analysis and this can be accomplished by evaporation.

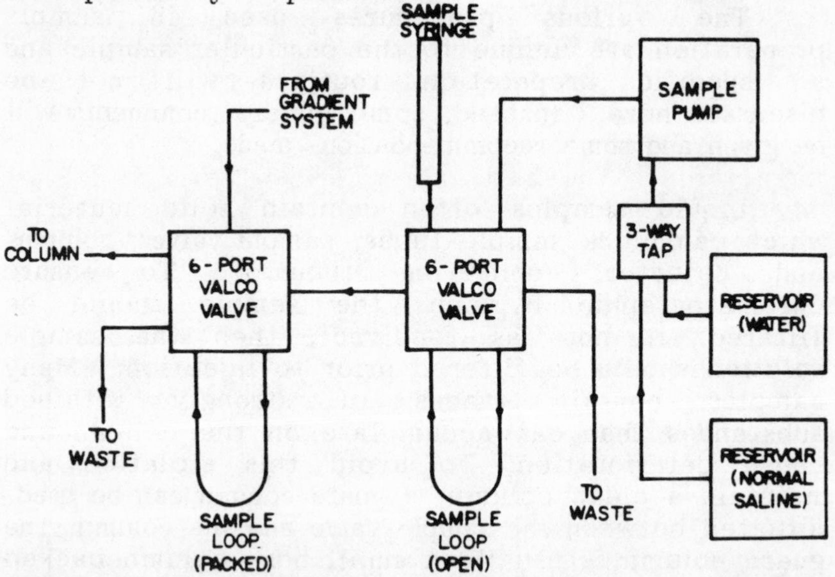

Figure 4. Block Diagram of Valve System for Sample Concentration.

Care should be taken not to overheat the sample to ensure that no thermal decomposition takes place.

Furthermore, the surface of the sample should have a current of inert gas passed over it, such as nitrogen or argon, to prevent oxidation during the evaporation procedure. An alternative method of sample concentration, that is particularly useful if the materials of interest are present in the sample at very low concentrations, is by chromatographic adsorption (13,14). An apparatus suitable for this procedure is shown in Figure 4. The apparatus consists of two six-port valves and a small pump constructed of inert material. The first valve controls the sample access to an open sample loop, the second to a sample loop packed with reverse phase material having a large particle diameter to reduce sample flow impedance. The pump can draw its supply from either of two reservoirs containing solvents suitable for the sample, for example, normal saline and water. The sample pump can displace the sample in the open loop through the packed loop and allow it to be washed with either water or saline. The sample concentration procedure is as follows. A syringe is used to place the sample in the open loop. The valves are rotated and the sample displaced by the pump through the packed loop and out to waste. The substances of interest are adsorbed on the reverse phase at the beginning of the packed loop. After washing well with water, the valves are again rotated and the mobile phase passed in the reverse direction through the packed sample tube lifting the solute off the packing and onto the column. An example of the use of this method of sample concentration is shown in Figure 5 where a solute acetophenone is extracted directly from an aqueous solution. It is seen that the concentration procedure is very effective and allows a sample of 10 ml to be used providing a discernable peak for acetophenone at a concentration level of 1 ppb.

It is essential to employ extremely pure solvents in sample preparation as it is for the mobile phase.

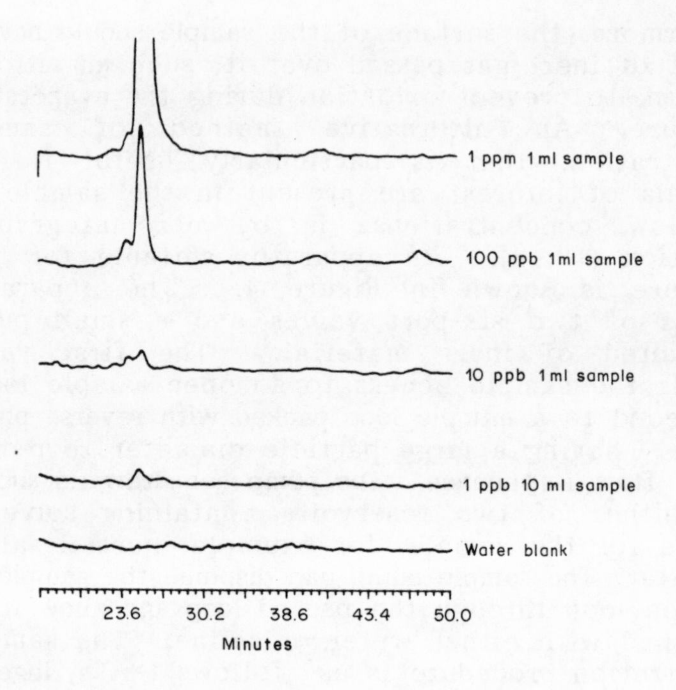

Figure 5. Chromatograms Demonstrating the Sample Concentration System. Column: 50 cm x 1 mm; Solvent: 75% v/v MEOH/25% v/v Water; Packing: 0DS2 10 micron; Flow Rate: 40µl/min; Sample: Acetophenone.

All solvents should be 'distilled in glass', and water, if used, should be cleaned by passing through a reversed phase column. Water purified in this manner is commerically available. The solvent used for sample injection should be the same as the mobile phase employed with the column, wherever possible. If the sample solvent is different or does not have the same composition as the mobile phase, spurious peaks will appear on the chromatogram and may be incorrectly interpreted as sample components.

The maximum sample volume that can be employed in any analysis will depend on the nature of the sample and the column used. If the sample has to be dissolved in a solvent other than the mobile phase, then the sample volume must be as small as possible to minimize any spurious peaks that may appear. If the sample is dissolved in the mobile phase, then the maximum sample volume should be used to allow the detector to be attenuated and thus provide minimum noise and drift.

The maximum sample volume that can be used is given by the following equation:

$$V_i = 2.6 r_c^2 \, l/(n)^{1/2} \qquad (4)$$

where V_i is the maximum sample volume,
r_c is the column radius,
l is the column length
and n is the column efficiency.

Injection Procedures

Today the LC sample is placed on the column exclusively by means of a sample valve, the days of syringes and high pressure septums are long since past. However, the hypodermic syringe is still used to place the sample in the sample valve prior to injection. The most precise and accurate results are obtained by the use of automatic samplers. Such sampling devices can have a capacity for as many as 100 sample vials. Used in conjunction with a laboratory robot to prepare the sample, the whole system can be completely automatic and, in fact, has been shown to provide the most accurate and precise quantitative results. Furthermore, because the robot is re-programmable or can be programmed to carry out a number of quite different sample preparation procedures, the robot-LC combination is an extremely versatile analytical system. An overview of laboratory automation procedures is given in Chapter 8. For further information regarding the use of robots the reader is referred to a book to be published in this series edited by Lochmuller entitled "Robots in the Laboratory" (15).

Irrespective of whether the sample is manually or automatically injected, the sample valve must be designed correctly if a reasonable life is to be realized from both the valve and the associated column. In general the life of the valve is extended considerably if it is not operated at its maximum working pressure. As a general rule, a sample valve should not be operated continuously over a long period of time at pressures much greater than two thirds the maximum rated. Column life will also be seriously impaired if a simple in-line sample valve is employed. If the mobile phase flow is stopped during the rotation of the valve, when the flow is again started the full pressure of the pump is suddenly applied to the packing causing its disruption. This effect can destroy a column in a few days and sometimes in a few hours. By-pass valves should always be employed for sample injection. These valves are designed such that the pressure applied to the column is constant throughout the whole injection procedure. The valve is in fact analogous to the 'make-before-break' electronic switch.

One final point on the subject of sample valves and this again refers to chromatographic hygiene. Whether manual or automatic sampling is employed, the valve syringe and any other associated equipment must be well washed with appropriate solvents between samples. Sample carry-over is one of the most common sources of error in quantitative LC.

Operating Conditions

Accurate and precise quantitative analyses can only be obtained in LC if the pertinent chromatographic variables are maintained sufficiently constant. The control of some of these variables is not in the hands of the chromatographer. The constancy of the mobile phase flow rate provided by the pump, the tolerance of the sample valve to high pressures, the quality of the column packing and the linearity of the detector are properties built into the equipment during manufacture and consequently cannot be changed. It follows that it is extremely important to be confident of the performance of the chromatograph or the chromatography modules before purchase. The necessary specifications will be available from all reputable manufacturers and if not, perhaps an alternative supplier should be sought.

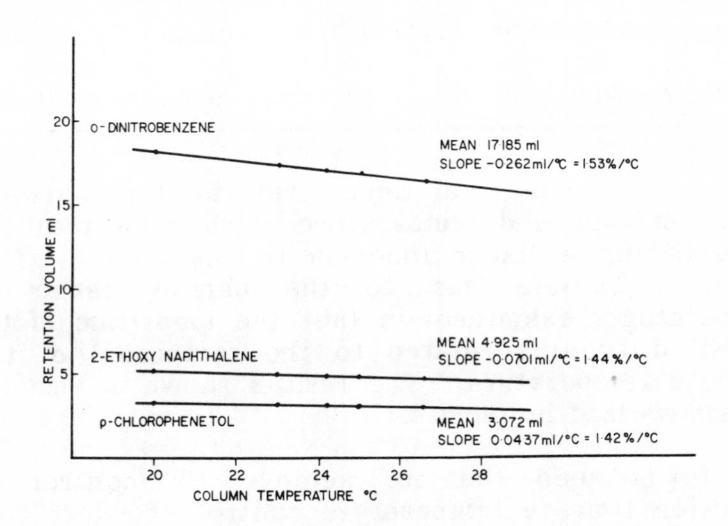

Figure 6. Graphs of Corrected Retention Volume Against Column Temperature for Three Solutes.

There are, however, some variables that are under the control of the chromatographer that must be carefully controlled and these are operating temperature, solvent composition and sample mass. The necessary control of these variables was investigated by Scott and Reese (6).

The *effect* of *temperature* on retention time is shown in Figure 6.

TABLE 4

Temperature Tolerances for Retention Precision

Solute	t_R (min) at 28°C	k'	Temp. Control for 1% Precision (°C)	Temp. Control for 0.1% Precision (°C)
ρ-chlorophenetole	3.07	0.945	±0.35	±0.04
2-ethoxy naphthalene	4.93	1.519	±0.35	±0.04
o-dinitrobenzene	17.19	5.301	±0.33	±0.03

The curves depict a linear relationship between retention time and temperature which is the result of force-fitting a linear function to the data. This is only possible due to the narrow range of temperatures examined; in fact the logarithm of the retention time is related to the reciprocal of the absolute temperature. The results shown in Figure 6 are summarized in Table 4.

It is seen that to achieve 1% control of retention time, a temperature control of ± 0.35°C is necessary. If a 0.1% control of retention time is required then the temperature control must be ±0.04°C. This level of temperature control is impossible without careful thermostating. It should be pointed out that, although such control is possible with a high heat content thermostating medium, it would be almost impossible to reset the temperature to this accuracy after prior change. It follows that if retention times are to be measured with a 0.1% precision, the column must be recalibrated if the thermostat has to be reset.

The effect of *solvent composition* on solute

retention is shown in Figure 7. Again due to the small concentration range over which the measurements were carried out, the results could again be force-fitted to a linear curve. The results are summarized in Table 5.

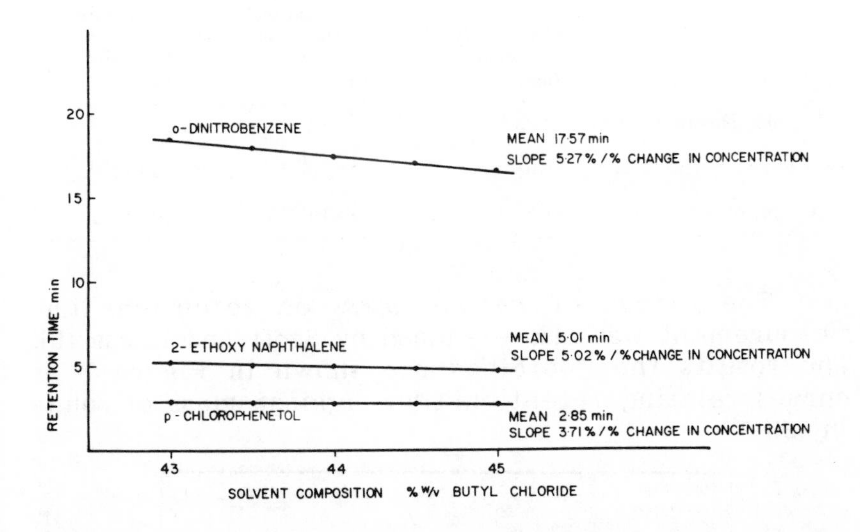

Figure 7. Graphs of Corrected Retention Time Against Solvent Composition for Three Solutes

It is seen that the necessary control over solvent composition to ensure precise retention measurements can be even more demanding than temperature control and in fact more difficult. For 1% precision the solvent composition must be kept constant to within 0.1%. For 0.1% precision (which as will be seen later can, under some circumstances, be extremely useful) the solvent concentration tolerance is 0.01%. It is impossible to make mixtures of volatile solvents to this accuracy under normal laboratory conditions. Consequently, to ensure the required precision, the mobile phase should be made up in bulk and the system recalibrated each time a new batch is used. The mobile phase reservoir should also be closed to atmosphere to eliminate evaporation losses.

TABLE 5

Solvent Concentration Tolerances for Retention Time Precision

Solute	Retention Time at 44% w/v BuCl in n-heptane (min)	Concentration tolerance for 1% Precision (% w/v)	Concentration tolerance for 0.1% Precision (%w/v)
ρ-chlorophenetol	2.85	±0.14	±0.015
2-ethoxy naphthalene	5.01	±0.10	±0.010
o-dinitrobenzene	17.27	±0.10	±0.010

The effect of *sample mass* on retention time measurement was also examined by Scott and Reese (6). The results they obtained are shown in Figure 8 as curves relating retention time against mass of solute injected.

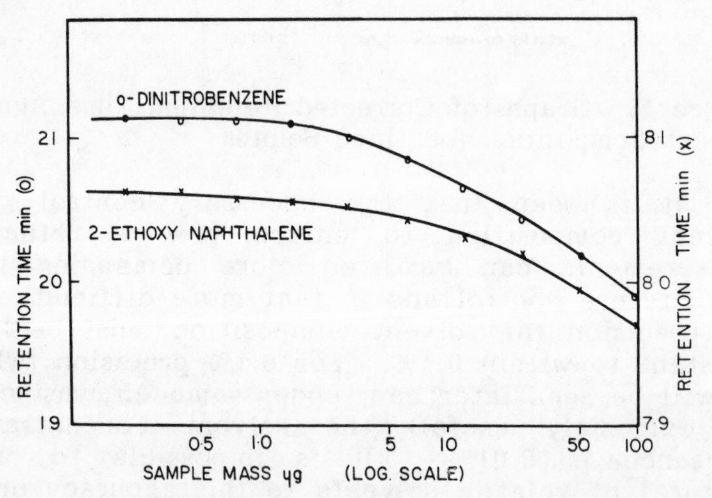

Figure 8. Graphs of Retention Time Against Sample Mass for Two Different Solutes.

It is seen that retention times are indeed

significantly effected by the mass of sample and consequently for precise results the same mass of sample must be placed on the column for replicate analyses. Furthermore, the mass of sample must be below a minimum value depending on the column dimensions. The results shown in Figure 8 are from a column 25 cm long, 4.6 mm in diameter, packed with 10 micron silica gel. The results indicate that the maximum mass of any solute must not exceed one microgram if precise retention data is required. This value will change with the column dimensions, and shorter columns of smaller diameter will have a smaller maximum sample mass that can be employed before effecting the measured retention time. It follows that the sensitivity required from the detector will also be greater for smaller diameter, shorter columns.

QUANTITATIVE ANALYSIS

There are three basic quantitative analytical methods that can be employed in LC. They are the *Internal Standard Method*, the *External Standard Method* and the *Normalization Method*. These methods have various and different attributes. The internal standard method is more tedious as it requires a known amount of standard to be added to each sample and the standard must be completely resolved from all the components of the sample. It does, however, provide the greatest accuracy and precision.

The external standard method requires a separate calibration chromatogram to be run, but only one standard solution needs to be made up in bulk and can be used for a large number of analyses. It also has the advantage that the actual solutes of interest themselves can be used as standards and thus allow the peaks to be normalized between the sample and the reference chromatograms. In general, however, this method is less precise and less accurate than the internal standard method because the data from at least two chromatograms are required for each analysis.

The normalization method is the simplest and can provide the most accurate and precise results. Unfortunately it can only be used in very special circumstances where the response of the detector is the same for all solutes of interest. In this method each peak area or peak height is expressed as a percentage of the total peak areas or total peak heights. One example where a normalization procedure can be employed has already been mentioned and that is in the use of the RI detector for analyzing polymer mixtures.

The Internal Standard Method

Firstly, a substance that is completely separated from all the solutes in the sample is selected as the standard. Then a synthetic mixture is made up of known weights of the standard and all the solutes of interest. In the example given below, it will be assumed for algebraic convenience, that there is only one solute to be determined quantitatively in the mixture.

Let there be m_X gram of the solute to be determined and m_{st} gram of the standard in the calibration mixture. Let the peak heights for the solute of interest and the standard be h_X and h_{st} respectively.

$$\text{then} \quad m_X/m_{st} = \phi h_X/h_{st} \tag{5}$$
$$\text{or} \quad \phi = m_X h_{st}/m_{st} h_X$$

where ϕ is known as the calibration constant.

A known weight of the standard m'_{st} is now added to the sample and the mixture chromatographed. If h'_X and h'_{st} are the peak heights of the solute of interest and the standard, respectively, then $m'_x/m'_{st} = \phi\ h'_x/h'_{st}$ (6)

Substituting for ϕ from equation (5) the weight of the solute of interest in the sample is given by:

$$m'_x = (m_x h_{st}/m_{st} h_x)\ (h'_x m'_{st}/h'_{st}) \quad (7)$$

This equation is quite general and can be used for any number of solute peaks if required. Peak area values can be directly substituted for peak heights in equation (7) if so desired. The relative accuracy and precision that can be achieved by peak height or peak area measurements will be discussed later.

The External Standard Method

The external standard method requires the standard to be chromatographed separately from the mixture and consequently the chromatographic conditions must be maintained extremely constant. In almost all LC analyses that employ the external standard method, the standard reference compound or compounds are identical to the solute or solutes of interest. This is one of the advantages of external standards as no relative response factors need to be determined. The concentration of the standard in the reference mixture is usually made up to be about the same concentration as the solute in the sample mixture. Thus the peak height of the solute in the sample is close to that of the standard in the reference mixture and consequently, this reduces possible errors that might arise from slight detector non-linearity. The success of this method depends on the reproducibility of the volume of injection and consequently it is more often used with automatic injection devices.

The method of calculation is very simple. If the concentration of the solute in the reference solution is c_{st} and that in the sample c_x and the peak heights of the reference peak and the solute peak are h_{st} and h_x, respectively, the concentration of the solute in the sample is given by:

$$c_x = c_{st}h_x/h_{st} \qquad (8)$$

Equation (8) is also quite general and can be used for any number of solutes, and peak heights can be substituted for peak areas. It should be remembered, however, that if the reference solute and sample solute are not identical or one reference compound is used to determine more than one sample solute, then the appropriate response factors must be determined as in the case of the internal standard method.

For the maximum precision and accuracy, the standard and sample should be run alternately. Thus, two chromatograms have to be run for each analysis. If chromatographic conditions can be kept very constant, then standards can be run every 3 or perhaps even 6 samples but such a procedure runs the risk of reduced precision.

The Normalization Method

The normalization method is the simplest but unfortunately, the least likely to be appropriate for LC analyses. Very few samples have components that all have the same response characteristics with the detector being used and this is a prerequisite for this method to be employed. If there are 1,2,3,--n solutes in the mixture having peak heights, h_1, h_2, h_3,----h_n then the percentage of any solute P is given by:

$$\%P = 100h_p/(h_1+h_2+h_3 \text{-----}+h_n) \qquad (9)$$

Equation (9) is also general and can be applied

to any mixture with any number of components. Peak heights can be substituted for peak areas if so desired.

THE PRECISION OF MODERN LC ANALYSES

Scott and Reese (6) attempted to identify the precision that could be obtained from LC analyses employing contemporary chromatographic equipment. They used a Waters 6000A pump, a 1-µl Valco sampling valve and a fixed wavelength UV detector operating at 254 nm. The mobile phase supply passed through the column thermostat prior to entering the sample valve and column which were also situated in the thermostat. Temperature control was ±0.05°C. A computer data acquisition system was employed together with a passive filter circuit with a time constant of 0.45 sec to reduce noise. Data was acquired at a rate of 15 data samples per second and the computer threshold acceptance level was 0.3 µV.

TABLE 6

Precision of Retention Time Measurement in LC

	Peaks		
	1	2	3
k' value	0.94	1.50	5.2
		Retention Time	
Mean (min)	6.283	8.119	20.421
S.D (sec)	0.38	0.20	0.46
S.D % of the mean	0.10	0.04	0.04

Twelve replicate samples were analyzed and the results from the retention time measurements shown in

Table 6.

It is seen that the precision attained was quite remarkable. For the peak with a retention time of over 20 min the standard deviation was less than half of a second which was equivalent to 0.04%. Even for the solute with a retention time of just over 6 min the standard deviation was less than 0.4 sec equivalent to 0.10%.

TABLE 7

Precision of LC Analysis by Peak Area and Peak Height Measurements

k' value	0.94	1.50	5.21
Analysis by Peak Height			
Mean	1.937	16.491	81.574
S.D.	0.0465	0.121	0.148
S.D. (% of mean)	2.46	0.736	0.18
Analysis by Peak Area			
Mean	0.633	7.486	91.884
S.D.	0.032	0.072	0.0823
S.D. (% of mean)	5.071	0.97	0.09

The precision obtained for peak height and peak area measurements is shown in Table 7. It is seen that peak height measurements give better precision for the peak eluted at the lowest k' value but there is little to choose between the two methods for the peak eluted at a k' value of 1.5. At a k' of 5.21, however, peak area measurements provide a lower standard deviation than that by peak height measurements as the peak is much wider and consequently the area measured is much greater.

In general if the individual peaks in the sample are not adequately resolved, peak height measurements are to be preferred as the measurements are more simple and peak height values are not very sensitive to small changes in flow rate. Conversely if the peaks are well resolved, estimate of the individual peak areas is likely to be more precise than that of peak heights. Peak areas, however, will be far more sensitive to fluctuations in column flow rate.

Provided retention times can be measured with sufficiently high precision, then it is possible to measure the composition of a binary unresolved mixture from the retention time of the composite peak. This, however, is only possible if the standard deviation of the retention time measurement is small compared with the retention time difference of the individual solutes. An example of this procedure is as follows.

Consider two solutes eluted close together such that a single composite peak is produced. From the plate theory, the concentration profile of such a peak can be described by the following equation:

$$c_{AB} = \frac{c_A}{\sqrt{2\pi n_A}} \cdot \exp[-(v_A - n_A)^2 / 2n_A]$$

$$+ \frac{c_B}{\sqrt{2\pi n_B}} \cdot \exp[-(v_B - n_B)^2 / 2n_B] \quad (10)$$

where c_{AB} is the concentration of solutes A and B at any point in the composite peak, c_A is the initial concentration of A, c_B is the initial concentration of solute B, n_A is the column efficiency for solute A, n_B is the column efficiency for solute B, v_A is the volume of mobile phase passed through the column in units of plate volumes of solute A, and v_B is the volume of mobile phase passed through the column in units of plate volumes of solute B.

If t_A and t_B are the retention times of solutes A and B equation (10) can be transformed into:

$$c_{AB} = \frac{c_A}{\sqrt{2\pi n_A}} \cdot \exp[-(n_A/2)(t/t_A-1)^2]$$

$$+ \frac{c_B}{\sqrt{2\pi n_B}} \cdot \exp[-(n_B/2)(t/t_B-1)^2] \qquad (11)$$

where the variable (v) is now replaced by variable (t), the elapsed time. It is seen from equation (11), that when only solute A is present, the function will exhibit a maximum at $t = t_A$ and, if only solute B is present, it will exhibit a maximum at $t = t_B$. It follows that the composite curve will give a range of maxima between $t = t_A$ and $t = t_B$ for different ratios of c_A to c_B and thus from the value of t at the maxima of the composite peak, c_A/c_B can be determined.

For closely eluted peaks $n_A = n_B$ and thus, as the function $2n_A$ is in effect an average dilution factor resulting from the dispersion, they can be replaced by a constant. The efficiencies n_A and n_B in the exponent function, however, can only be considered equal if the peaks are symmetrical as, in the part of the composite peak that determines its maximum, the rear part of the first peak is combined with the front part of the second peak. In liquid-solid chromatography, the concentration profiles of eluted peaks are rarely symmetrical, and thus, n_A must represent the efficiency of the front half of solute B. Further, the detector response to solutes A and B must be taken into account. Thus, if D is the detector signal then equation (11) can be put into the form

$$D=\psi\{\alpha c_A \cdot \exp[-(n_A/2)(t/t_A-1)^2]+$$

$$\beta c_B \cdot \exp[-(n_B/2)(t/t_B-1)^2]\} \qquad (12)$$

where ψ is a constant,
α is the response factor of the detector to solute A
and β is the response factor of the detector to solute B.

The system was examined employing nitrobenzene and fully deuterated nitrobenzene as the solutes. Their elution times were 8.927 and 9.061 min, respectively, thus having a retention difference of 8.04 sec. The separation ratio of the two solutes was 1.023, and the efficiencies of the front and rear portions of the peaks were 5908 and 3670 theoretical plates, respectively. The detector was found to have the same response for both solutes, i.e. $\alpha = \beta$. Thus, inserting these values in equation (12):

$$D=\psi\{(c_A \cdot \exp[(-3670/2)(t/8.927-1)^2]$$

$$+c_B \cdot \exp[(-5908/2)(t/9.061-1)^2]\} \qquad (13)$$

Employing a range of values for c_A/c_B the retention time of the composite peak was calculated from equation (13) by means of a computer. The curve relating the composition of the mixture to retention time is shown in Figure 9.

A series of mixtures of nitrobenzene and deuterated nitrobenzene were made to a known concentration ratio and the retention times of the composite peaks were determined experimentally. The retention time of each mixture was determined in triplicate, and the average for each mixture is represented as plotted points in Figure 9.

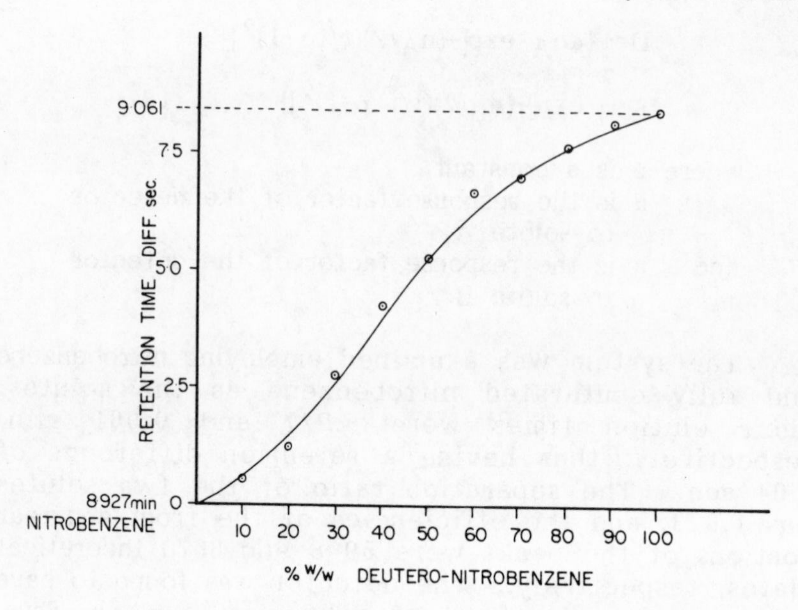

Figure 9. Curve Relating Composition of Mixture to Retention Time.

It is seen that close agreement is obtained between the experimental points and the theoretical curve.

Employing very precise methods of measuring retention times as a means of determining the composition of unresolvable binary solute mixtures would be extremely valuable in the analyses of configurational isomers. Providing the retention time of a known mixture of the two components is available (in most instances one pure isomer and a 50%(w/v) mixture of one isomer in the other can be obtained) a calibration curve can be calculated theoretically. The asymmetry ratio of the peak for one pure component and the column efficiency for that component is usually the only further information required as the detector response factors for configurational isomers are generally identical.

SYNOPSIS

The first priority for *accurate* and *precise quantitative LC* analysis is *adequate resolution*, sophisticated mathematical algorithms are not a satisfactory alternative. Resolution between solute peaks of three standard deviations is the minimum acceptable. Optimum ergometric layout of LC equipment coupled with *careful chromatographic hygiene* are prerequisites for *good quantitative precision*. There is an optimum column for any LC analysis but if this is not available the column should be run at the optimum velocity and be loaded with the optimum sample volume. *Baseline stability is essential* and to ensure this, column and detector *temperatures* and *flow rate* should be *kept constant*. Equilibrium with the mobile phase should be ensured, guard columns should be used and columns should be replaced when irreversibly contaminated. The most *widely used detectors* for quantitative LC are the *UV* detector, the *RI* detector, the *electrical conductivity* detector and the *fluorescence detector*; the first having the widest linear dynamic range, the latter having the smallest linear dynamic range but the highest sensitivity. The *UV and fluorescence detectors* can be employed with *gradient elution* if appropriate solvents are selected; the RI detector is unsuitable for gradient elution while the electrical conductivity detector can be employed for gradient elution if the ionic strength of the solvent is kept constant. Employing derivatization procedures the *limited linearity* of the *fluorescence detector* is often *tolerated* for the sake of its *high sensitivity*. The linearity of any detector employed for quantitative analysis should be known or determined, preferably by measuring the Response Index of the detector. The numerical value of the *Response Index* should be *between 0.98 and 1.02* if linearity is to be assumed. Most samples need some preparation which should be carried out with great care and with the best possible chromatographic hygiene. *Clean solvents* should always be used and all *samples should be filtered* before injection into the sample valves. Solid samples will need dissolution or extraction which must also be followed by filtration. The *maximum sample volume* employed will depend on the *column dimensions* and efficiency. If the sample preparation is complex the *laboratory robot* will provide *improved precision, economy* and sample *throughput*. The sample valve should not be employed continuously at its maximum rated pressure to ensure reasonable valve life. *By-pass valve systems* should be employed to *extend column life*. All sample valves should be cleaned well between samples. Accurate retention measurements (standard deviation 0.1% of the mean) require *temperature control to 0.04°C* and *solvent composition control to 0.1% v/v*. The mass of any individual solute injected should be kept below a maximum of 1 μg for high precision. There are three basic methods for quantitative analysis: the *Internal Standard Method*, the *External Standard Method* and the *Normalization Method*. The *Internal Standard* method requires the addition of a standard to each sample and a precalibration procedure to determine the response factor for each solute concerned. This procedure is tedious, time consuming but probably gives the *most precise results*. The *External Standard Method* requires only one calibration mixture to be made but needs reasonably frequent calibration runs. It can use the same solute for calibration as that in the mixture to be determined and thus *eliminates the need to determine response factors*; it probably gives *slightly less precise results* than the Internal Standard Method. The *Normalization Method* requires that the detector gives the same response for all the solutes in the mixture. It can be *rarely used* but when it is, it is usually with the RI detector; this method is capable of *providing the best precision* of all three procedures. In general peak area and peak height measurements give similar quantitative precision; if there is *a preference* it *might* be the use of *peak height measurements.* If retention times can be determined with very high precision, such measurements can be employed, under special circumstances, for the quantitative estimation of a pair of solutes that elute as a single peak.

References

1. E. Katz, K. Ogan and R.P.W. Scott, J. Chromatogr., 260 (1983) 277.

2. R.P.W. Scott in "Small Bore Liquid Chromatography Columns", R.P.W. Scott (Ed.), Wiley, New York, 1984, p. 17.

3. J.J. Van Deemter, F.J. Zuiderweg and A. Klinkenberg, Chem. Eng. Sci., 5 (1956) 271.

4. J.H. Knox and M. Saleem, J. Chromatogr. Sci., 7 (1969) 614.

5. E. Katz, K. L. Ogan and R.P.W. Scott, J. Chromatogr., 289 (1984) 65.

6. R.P.W. Scott and C.E. Reese, J. Chromatogr., 138 (1977) 283.

7. R.P.W. Scott, "Liquid Chromatography Detectors" Elsevier, Amsterdam, 1986, 235.

8. "Chemical Derivatization in Analytical Chemistry", R.W. Frei and J.F. Lawrence (Eds.), Plenum Press, New York, 1982.

9. F.M. Rabel, Amer. Lab., 6 (1974)33.

10. I.A. Fowlis and R.P.W. Scott, J. Chromatogr., 11 (1963) 1.

11. C.E. Reese, J. Chromatogr., 18 (1980) 201.

12. C.E. Reese, J. Chromatogr., 18 (1980) 249.

13. C.B. Eurton and D.R. Baker, Amer. Lab., March (1979) 91.

14. R.P.W. Scott and P. Kucera, J. Chromatogr., 185 (1979) 27.

15. "Robots in the Laboratory" (Ed. C.H. Lochmuller) Wiley, New York , London to be publisher.

Quantitative Analysis using
Chromatographic Techniques
Edited by Elena Katz

Chapter 4

DETECTION IN QUANTITATIVE GAS CHROMATOGRAPHY

Peter C. Uden

Any analytical separation process must be integrated with a means for its evaluation. In a simple situation, in which only an indication of the resolution efficacy is desired, a qualitative evaluation of the separation efficiency may suffice. Visual examination of a thin layer chromatogram exemplifies such a situation. However, when quantitative analysis is needed, or when the identification or characterization of eluted components is required, a dedicated instrumental detection system is mandatory. The essence of such detection is a device which is situated immediately at the end of a chromatography column to receive eluates with minimum delay and to respond to the presence of the solute in the mobile phase in a quantitatively predictable fashion.

Since gas chromatography (GC) was the first effective high resolution chromatographic separation process developed, devices to detect and quantitate dilute levels of vapors in flowing streams of carrier gas were the first to be examined. In contrast to static systems such as radiation detectors in spectroscopy, wherein a signal is generated by the radiation input at the detector, the GC detector must respond in 'real time' to an ever varying concentration of analyte molecules. This is achieved by continually monitoring an appropriate property of the system and displaying the resulting output in a suitable form, usually as a chromatogram on a chart recorder.

During the 30 years or more of GC development, a plethora of such 'flowing stream' detection devices have been developed, wherein an independent analytical 'instrumental transducer' is employed to generate a signal from GC eluates containing solute vapor. Many detectors are now only of historical interest but some have achieved general acceptance and new systems are continually being considered. A broad range of physical and chemical properties of substances contained in a permanent gas as a vapor has been employed as detection modes. Detection criteria are being continually revised in view of new demands such as capillary GC and eluate characterization needs. This chapter seeks to provide an overview of the quantitative aspects of GC detection, comparing and contrasting the properties and performance of different detection devices. Quantitative response is an essential feature of any candidate GC detector and a wide linear dynamic range is critical since the mass contained in a GC peak may range from milligrams to picograms. in magnitude. A fast response of both sensor and associated electronics is also vital, particularly in capillary GC where time constants of 50 ms or less are necessary to preclude loss of column efficiency. Short term and long term baseline instabilities must be minimized and both high sensitivity and low minimum detectable quantities (MDQ) are also desirable objectives. Other generally desirable features of detectors include ease of use, ruggedness and low cost. An ideal detection system will also provide useful qualitative information on analytes for characterization and identification. No single detector can contain all these desirable features since some properties are conflicting in information content, e.g. detectors may be selective or universal.

CLASSIFICATION OF GC DETECTORS

All GC detectors depend, for their utility, on the measurement of a parameter or property which differs between pure carrier gas and the eluent/carrier gas mixture. The requirement is, in essence, the quantitative measurement of such a property difference when the ratio of minor to major component in the binary analyte mixture may range from 1: 10^3 to 1: 10^{12}. This is a demanding task. Detection systems have been classified under a number of formats:

i.) detectors can be *universal (U)*, *selective (SL)* or *specific (SP)*.
ii) detectors can have a response mode which is *mass dependent (MS)* or *concentration dependent (CS)*.
iii) detectors fall into a number of phenomenon-based classes: *ionization*, *bulk physical property*, *optical/spectroscopic*, *electrochemical* and *reaction based*.

Key to abreviations

TCD - Thermal Conductivity Detector
GDB - Gas Density Balance
USD - Ultrasonic Detector
FID - Flame Ionization Detector
HAFID- Hydrogen Atmosphere Flame Ionization Detector
PID - Photoionization Detector
TID - Thermionic Ionization Detector
HID - Helium Ionization Detector
ECD - Electron Capture Detector
FPD - Flame Photometric Detector
MS - Mass Spectrometer
IR - Infrared Spectrometer
AES - Atomic Emission Spectrometer
AAS - Atomic Absorption Spectrometer
HECD- Hall Electrolytic Conductivity Detector

Any detector will be characterized by one factor from each of these classifications; certain detectors however, function in more than one depending on operational circumstances. Table I lists a number of GC detection systems classified according to this scheme. Some of the quantitative characteristics of detectors from these different classifications will now be considered.

TABLE I

CLASSIFICATION OF GC DETECTORS

Class of Detection Phenomenon	Detector	Universal, Selective of Specific	Mass or Concentration Responsive
Physical Bulk Property	TCD	(U)	CS
	GDB	(U)	CS
	USD	(U)	CS
Ionization	FID	(U)	MS
	HAFID	SL	MS
	PID	SL	MS
	TID	SL	MS
	HID	SL	MS
	ECD	SL	MS/CS
Spectroscopic	FPD	SL	MS
	Chemiluminescence	SL	MS
	MS	U/SL	MS
	IRS	U/SL	CS
	AES	SL/SP	MS
	AAS	SP	MS
Electrochemical	Coulometric	SL	CS
	HECD	SL	CS
Reaction Based	Chemiluminescence	SL	MS
	HECD	SL	MS
	Methanizer FID	SL	MS

Mass Sensitive Detectors

Mass flow sensitive detectors are those in which the response is proportional to the mass passing through the detector in unit time. The performance can be defined in terms of the nature of the *response factor* or *sensitivity*. The definition best used is the magnitude of the signal observed for the sample relative to the mass injected, i.e. the ratio of signal to sample size. When the signal is measured as the peak height, the sample size is the mass flowrate through the detector at the peak maximum. The equation for the response factor is:

$$R = hw/2m$$

where m is the weight of the analyte injected,
w is the peak width
and h is the peak height.

In terms of the peak area (*A*), the response factor is simply given by *A/m* (here *hw/2* is taken as equal to *A*) Peak heights may be usefully expressed in terms of electrical response factors of the detector electrometer, e.g. 'x' nA full scale. *Detectability* (alternatively sensitivity, MDM) represents a certain minimum detectable mass (MDM) presented to the detector within a unit time (usually taken as twice or three times the detector noise level); for the flame ionization detector (FID) this may be *ca* 10^{-12}g/s. A key operational feature of mass flow rate dependent detectors is that their *response factors do not vary with carrier gas flow rate.*

Concentration Sensitive Detectors

For such detectors, the sample size is considered as the concentration in the detector at the peak maximum. The response factor equation is:

$$R = hwQ/2m$$

where Q is the flow rate (R = AQ/m in terms of peak area).

This type of detector, exemplified by the thermal conductivity detector (TCD) responds to a change in the concentration of the analyte substance in the carrier gas within the detector rather than to the presence of the substance itself, i.e. no response is obtained unless the composition of the flowing gas mixture changes. Since the output of this type of detector is typically reported as a voltage, peak areas can be expressed in volt-seconds and the response factor can be expressed in V/mass/volume.

Universal Detection

At first sight universal detection is a desirable goal; however, in practice it is often necessary to discriminate in favor of trace level components in the presence of a great excess of the matrix species to simplify the resolution problem. In this case universality may, in fact, constitute a disadvantage. For a GC detector to be truly *universal*, it must respond to any GC eluate. Only a few such detectors exist, primarily in the *bulk physical property* category in which any eluate modifies a measured property of the carrier gas. Only if the eluate property is identical with that of the carrier gas will there be no response. However, responses will naturally differ for each solute by virtue of a graduation in the magnitude of the observed parameter. No detector exists in which all analytes will respond equally although, in a restricted case such as where a flame ionization detector is used for petroleum hydrocarbon analysis, a *near-universal* response to all components may be achieved. Spectral property detectors can show universality in that all eluate molecules will exhibit a mass spectrum, infrared spectrum etc. The closest to the universal detectors may be the 'total ion monitoring' mode in GC - mass spectrometry or an integrated total spectral response in GC-vapor phase infrared spectroscopy.

Selective Detection

Detectors in which some particular property of the solute molecules is measured will show selectivity for those exhibiting that property. Since in virtually all cases, the measured property also appears, albeit to a much lesser degree, in other types of analyte, a sufficiently high concentration of such substances will also elicit a response and cause interference. The *selectivity* of the detector relative to this interference will determine its effectiveness. Detectors may be element selective, structure or functionality selective or *property* selective.

Specific Detection

Specific detectors are selective detectors that exhibit a very high degree of selectivity. High spectral resolution, or specific reactivity in reaction detectors, may cause detection to approach true specificity. High resolution mass spectrometry with multiple specific ion monitoring is probably the closest technique to absolute specific detection, although high resolution emission spectroscopy for element specific detectors is also in contention.

There is some confusion among chromatographers between selectivity and specificity. Both have been defined in terms of the amount of hydrocarbon that gives the same response as the analyte in terms of molecular weights or weights or elements. Some detectors may be highly selective but not very predictable in response to individual species, the electron capture detector (ECD) is such a detector.

CLASSIFICATION BASED ON THE DETECTION PROCESS

Ionization Detectors

Gas chromatographic carrier gases (with the

possible exception of the noble gases used under certain circumstances) behave as almost perfect insulators at normal temperatures and pressures. Any phenomenon induced in the eluate molecules which introduces electrical conductivity into the flowing gas stream permits detection. High impedence electrical circuits permit the transduction of current carrying effects into a displayed voltage signal. The various methods of ionization form the backbone of present day GC detection, the major techniques being flame ionization, photoionization, various modes of thermionic ionization, inert gas mediated ionization and electron capture detection.

In each type, a different phenomenon is used to generate the ion current but this is always aimed to be related quantitatively to the eluate properties. Ionization detectors can usually be considered as *solute property* detectors with respect to the carrier gas mobile phase since the ion current measured is zero or minimal in the absence of solute. Signals from such detectors (GC or LC) can usually be amplified electronically to a greater degree than those of *bulk property* detectors, thus generating intrinsically superior detecting limits (i.e., the signal to background is higher in the solute property detectors). Selectivity varies considerably among ionization detectors ranging from the almost universal flame ionization detector to the highly selective electron capture detector. Response factors may also vary widely among eluates which will be dependent upon the ionization mechanism.

Bulk Physical Property Detectors

As its designation indicates, this type of detector responds to some physical property of the bulk mobile phase GC eluent (i.e. carrier gas and eluted species). A change in this property level from that of the pure carrier gas, when sample is present, is then detected. Thus, a typical situation is when some vapor phase property such as thermal conductivity is measured and the output for pure carrier gas gives the 'baseline' value which corresonds to the detector 'zero'. Such detectors are also typically *concentration sensitive* as noted above. Of necessity the *relative* change in the property of the gas phase being measured is small since the actual analyte concentration in the carrier gas is low. Thus the detector must always measure a relatively small signal on a large background response. 'Bridge' compensation methods are typically used whereby the carrier gas property measurement is 'nulled out'; nevertheless absolute detection limits observed for these detectors are usually significantly poorer than for the ionization detectors.

Spectroscopic Detection

Vapor phase spectroscopic measurements are clearly attractive as GC detection modes for a number of reasons. The range of spectroscopies which may be considered suggests that universal detection may be approached or even achieved. If detection is not limited to a single spectral region but rather is integrated over a wide range, this is certainly true. Alternatively, specific detection may also be achieved by employing high spectral resolution. Additionally, sophisticated and sensitive instrumentation is independently available for various classes of spectroscopy. The disadvantage, however, is the high expense of most spectroscopic equipment relatively to the GC itself. Frequently, the spectral system is much more complex than the GC, furthermore, complicated interfacing between the two devices may be needed with resulting problems in quantitative sample transfer and maintenance of chromatographic resolution during detection. The advantages of GC/Spectroscopic detection have now established the family of techniques which include interfaced mass spectroscopy (GC/MS), infrared spectroscopy (GC/IR) and atomic emission spectroscopy (GC/AES). Quantitative aspects of all these and other spectroscopic detectors depend on spectral features and detection methods as much as on chromatographic separation.

Electrochemical Detectors

The primary group of GC detectors, which do not function directly as vapor phase monitors, are the *solution electrochemical detectors*. In practical GC systems, the eluted sample must be transfered into solution prior to detection. This additional step has clear implications both for quantitation and for maintenance of eluate peak integrity and resolution. A number of solution electrochemical parameters have been developed for GC detection, but relatively few organic compounds amenable to GC separation are electrochemically active with respect to oxidation or reduction processes in solution, or have noteworthy solution conductivity. Sometimes electrochemical selectivity is a valuable parameter for detection, but more often prior chemical or physical decomposition or transformation processes are needed to permit the eluate to respond electrochemically. The intermediate process must be carefully controlled or quantitative precision and accuracy will be compromised. Electrochemical detectors remain specialized detectors tailored for particular applications.

Reaction Based Detectors

A number of detectors with otherwise differing characteristics may be also considered together as *reaction detectors* wherein prior to detection the eluate is chemically or physically transformed into the actual analyte species. The electrochemical detectors which involve the pre-reaction mode form such a class. Pyrolytic or thermal degradation reactions such as are seen in the 'nitroso functionality' specific 'thermal energy analyser' also represent this approach. Catalytic transformation of eluates into species detectable by conventional detectors such as the FID again fall into this class. These detectors are highly selective in nature by virtue of reaction specificity. They are usually mass sensitive rather than concentration sensitive.

GENERAL QUANTITATIVE ASPECTS OF GC DETECTORS

The conventional detector output familar to all chromatographers is an analog signal providing response with respect to time. As already discussed in Chapter 2, computerized data systems operate through analog-to-digital conversion (A to D) and may display detection data, either directly from digital manipulation, or through subsequent digital-to-analog reconversion (D to A). In a complex system such as GC/MS, displayed chromatograms based on total or specific ion monitoring depend on the analog display of digitally stored data. This data must be obtained at acquisition rates commensurate with the demands of accurate quantitation and definitive resolution.

Baseline Stability

The establishment of a stable baseline for the duration of a chromatogram and its long term maintenance is a primary chromatographic objective. Ideally this baseline should show negligible signal from the presence of pure mobile phase carrier gas. Furthermore, the detector signal should not change with variations in flow rate, or depend on detector environment such as temperature or sample introduction procedures. If such system stability can be achieved, then detection measurements will be limited solely by the quality of the sensor and electrical circuit.

Signal-to-Noise Ratio

This measurement is common to virtually all analytical instrumentation and provide a means of determining the limiting detection level for a given solute. In the case of GC, this measurement is subject to the added complexity that results from the combination of electrical (amplifier) and system (column and sample system) effects. A typical portion of a gas chromatogram obtained at a maximum detector output setting is shown in Figure 1.

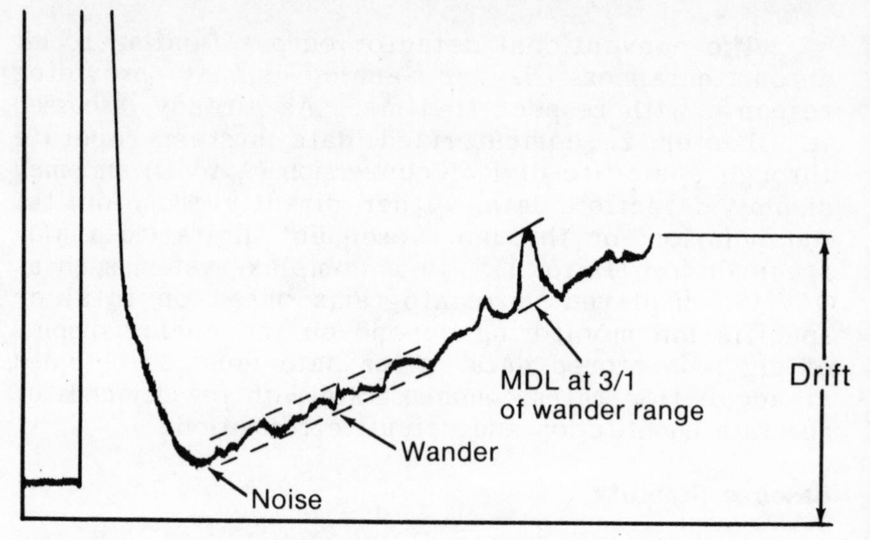

Figure 1. Factors in baseline stability.

Four features of quantitative importance are illustrated; *short-term noise*, *'signal-wander' (long-term noise)*, *drift* and *minimum detectable level (MDL)*. By spectroscopic analogy, long term noise and drift would be considered in 'signal to background'.

Short-term noise is noise having a frequency much greater than the measured event and can usually be eliminated by appropriate filters. Short-term noise is sometimes referred to as 'grass'. *Long-term noise* or 'wander' are signal excursions that take place over a time period similar to that of the measured event, here the GC peak. It is taken as the average distance between the highest and lowest excursions of the baseline during one peak width. In practice a section of baseline corresponding to several peak widths provides a better average measure of long-term noise (see Chapter 2). The defining lines for noise measurement are drawn just below the few highest long-term maxima and just above the few lowest long-term minima; measurement must be made such that the recorder responds adequately at the particular setting, i.e. it is not subject to 'deadband' limitation.

Drift is a directional change of the baseline occurring during long time periods relative to the total length of the chromatogram. While drift makes quantitative measurements difficult, it is typically controllable as it is often due to operational instrument parameters. Temperature changes of injector, column or detector or gradual changes in mobile phase concentrations such as those due to septum bleed are often responsible. Drift may also occur in high sensitivity detectors such as the ECD as a result of carrier gas purifier saturation. Drift is usually not considered in the calculation of the signal-to-noise ratio since its time scale is long in comparison to the measured chromatographic event. At high sensitivity settings long-term noise or 'signal wander' is often observed. This involves random baseline excursions having the same period as the peak width. It can arise for example from leaks in column fittings, plasma instabilities in spectral detection or mains fluctuations etc.

The defined signal-to-noise is thus subject to different interpretations and the total background noise signal referenced may change in different experiments. It must be remembered also that the chromatographic peak shape affects the usefulness of any defined S/N and any resolution changes will also directly change the S/N ratio.

Minimum Detectable Level

This form of detector specification is independent of column parameters (which influence resolution) or of actual sample size. The *minimum detectable level* is the sample concentration or mass in the detector at the peak maximum, when the S/N is at an agreed ratio. This ratio may be between 1 and 5 but the most widely used are S/N of 3 and 2. The MDL is sometimes called *detectability*.

The units of MDL depend on the detector class, i.e. whether it is a mass or concentration sensitivity detector.In a mass flowrate detector such as the FID, the MDL is obtained from peak half-width measurements. If the peak width at half-height is 5 sec for a sample of 1 ng per peak and the S/N value is 30, the MDL is given 1/(5x10) ng/sec for a defined MDL S/N of 3. Thus the MDL is 20 pg/sec. Mass/unit time are the dimensions employed for defining the MDL of mass flowrate responsive detectors.

For concentration responsive detectors a different situation applies. The MDL in mass/time units can be measured from the detector output chromatogram in the same way as for the mass flowrate sensitive detectors. Since these detectors respond to concentration changes, mobile phase volume corresponding to peak width time (and therefore mobile phase flow rate) must be taken into account. If, as in the above sample, the measured MDL of a peak is 20 pg/sec., but the carrier gas flow rate is 1 mL/sec, then the MDL based on concentration level becomes 20 pg/mL. A different column having a different flow rate (or the same column under different conditions) of 0.5 mL/sec would give an MDL based on mass flowrate of twice that noted before, i.e. 10 pg/sec. This is because, while the same mass flowrate of compound is passing through the detector, there is less carrier gas diluting it; thus a higher concentration sensitivity would be observed (based on mass/time units). For concentration responsive detectors then, MDL based on the mass/volume level is preferred since it is independent of carrier gas flow. In the above case, the MDL remains at 20 pg/mL for both carrier gas flow rates (at the slower flow MDL is 10 pg/sec./0.5 mL/sec. = 20 pg/mL).

There are alternative ways of expressing MDL for both mass flowrate and concentration responsive detectors. MDL may be given in terms of moles, i.e. moles/sec. or moles/mL. In the case of selective detectors, such as the element specific atomic emission spectrometric detector, MDL is expressed directly in terms of the weights of the elements detected, e.g. g(element) sec. MDL is often considered equivalent to *response factor*, however, it must be remembered that the former is calculated from a single point measurement while the latter is defined as the slope of the analytical detector response curve. Indeed, the response factor has been defined either as ratio of signal to sample size (then 'high sensitivity' defines the best performance) or sample size to signal (in which case 'low sensitivity' defines the best performance). For GC, this confusion is avoided by use of the MDL. If the 'linearity' of the detector is also known, then these two properties adequately define the detector behavior.

Linear Dynamic Range

Linear dynamic range is often confused with response factor, but is best defined in terms of signal response relative to the noise level. Perry has defined the linear dynamic range in GC detection terms (1). For each minimal unit in detector measured parameter (concentration or mass flowrate) there is a corresponding minimal unit increase in signal. The number that describes the linear dynamic range is the number of these corresponding minimal units. A plot of the concentration or mass per unit time (effectively mass) on the abscissa versus the signal on the ordinate will be a straight line passing through the origin and extending over the linear dynamic range. Alternatively expressed, the ratio of the concentration or the mass flowrate change to the resultant signal is invariant over the linear dynamic range. Eventually, another unit increase in measured parameter will not produce a corresponding unit increase in signal; the response has now passed outside the linear dynamic range.

TABLE II

CHARACTERISTICS OF SOME WIDELY USED GC DETECTORS

Detector	MDL	Linear Dynamic Range
TCD	3 x 10^{-9} g/ml	10^4
FID	10^{-12} g/s	10^6
AFID(NPD)	1 x 10^{-13} g/s (N) 1 x 10^{-14} g/s (P)	10^4
ECD	10^{-13} g/mL (lindane)	10^4 (pulsed frequency)
FPD	2 x 10^{-11} g/s (S) 10^{-12} g/s (P)	10^3 (S) 10^5 (P)
HECD	10^{-12} g/s (N) 5 x 10^{-13} g/s (S) 10^{-13} g/s (C1)	10^4 - 10^6

This limit is usually defined by a certain percent

departure from linear response, often 5%.

Linear dynamic range does not, in practice, depend only upon the GC detector. Other features of the chromatographic system are also pertinent. General ranges are quoted for all detectors however and these are valuable for practical analysis.

Typical values of MDL (for specific compounds) and general linear dynamic ranges are shown in Table II for a number of the more commonly used detectors.

Some other operational features of a GC detector linearity should be considered. The 'allowable' deviation from linearity needs to be considered when discussing the smallest 'linear' level. As O'Brian has commented (2) "What is the smallest 'linear' level? Is it the MDL, or is it a level sufficiently higher than the noise to allow a measurement to be made with the precision equivalent to the allowable deviation from linearity? Understandably, instrument manufacturers prefer the former definition, chromatographers with the responsibility of specifying the accuracy of their analyses prefer the latter".

Linear dynamic range can always be optimized by using experimental conditions which, however, may not be always practical. Peak area measurements remain linear over a wider range than peak heights, best linearity holds for low temperatures and, overall, for low boiling compounds in a gas sample. Rapidly eluting narrow peaks give best results. Lengthy instrument equilibration is almost always necessary. As a good operational guide, practical linear dynamic ranges up to one order of magnitude less than optimal may be expected.

Plots of the logarithm of detector response versus logarithm of sample sizes are widely used but, unless the slope of the curve is taken into account (which should be unity for a linear detector) such plots optimize the appearance of linearity (see Chapter 1 and 2). A more informative method of showing linearity is given in Figure 2 (3).

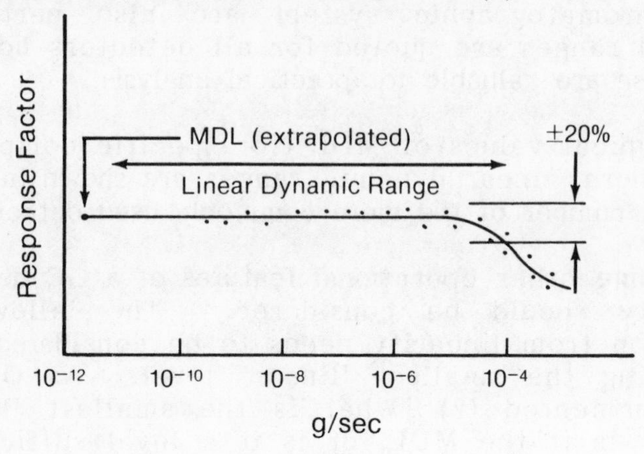

Figure 2. Linearity plot for FID.

Response factor is plotted on the ordinate and concentration, mass flow rate (or mass) on the abscissa. The curve is for a FID with a linear dynamic range of about 7 orders of magnitude as indicated by a straight horizontal line. Care must always be taken when comparing both MDL and linear dynamic range limits to note the operating parameters that are employed in their measurement.

Standardization

Universal standardization is not possible for any GC detector. There are no detectors with completely predictable absolute response factors although some, such as the FID, have responses for most compounds within a factor of 2 or 3. In these cases the scale of a chromatogram is predictable although quantitation cannot be obtained accurately from known references for unknown components. Calibration and standardization is thus necessary for quantitation of known analytes or of others that are chemically similar to them. For some highly selective detectors, such as the ECD, standardization for each analyte is essential for any quantitation to be valid. Selectivity of response has been mentioned earlier and this affects quantitative standardization directly in terms of detection intereference. In addition to providing a need to determine the signal-to-noise ratio, selective response can modify the background signal directly and thus can also alter the signal intensity. These interferences may result in either a reduction in response in the presence of co-eluents or unresolved background material, or conversely, in enhancement. An example of the former is the diminished response of the flame photometric detector for sulfur and phosphorus in the presence of a large excess of co-eluting hydrocarbons. Exemplifying the enhancement effect is the HAFID (hydrogen atmosphere flame ionization detector); in this case the effect produces a real sensitivity enhancement of analytical value.

To overcome these problems, a number of different sample analyses are needed. A sample blank is required comprising a mixture similar to the unknown; a standard mixture of known concentrations of the target analytes; the unknown mixture itself, and ideally a 'spiked sample' being a real 'unknown' to which known standard is added.

DETECTION SYSTEMS AND THE INTEGRAL GAS CHROMATOGRAPH

The ideal detector responds solely to column eluates and is independent of other operational factors. However, in reality, the overall chromatograph configuration and its operational parameters impact to a great extent on detection characteristics. Among the principal factors which need to be considered are: i) column type, whether preparative or analytical,packed or capillary; ii) the column - detector interface; iii) the system operating parameters; iv) multiple detector combinations and configurations; v) sample introduction; and vi) data retrieval, supporting electronics and computation facility.

The impact of the total system on *detector quantitation* will now be considered.

Packed and Open Tubular Columns

The rapid acceptance of high resolution capillary columns for many analytical purposes, (prompted particularly by the advent of fused silica columns) and the parallel diminution in the use of packed columns have placed increased demands on detector design and quantitative performance. Yang and Cram (4) have pointed out some principal considerations. Packed column GC rarely probes the performance limits of most detectors (5), but the inherently smaller sample levels and peak quantities necessary for efficient capillary separations, (typically a few micrograms or less) call for detectors with high baseline stability. Recently, miniaturized detectors with sufficiently low internal volumes and rapid response characteristics have been introduced, specially designed for capillary applications so as not to compromise column performance.

In capillary GC, the small quantity of material in each solute zone, the small peak width obtained by rapid elution, low carrier gas volume flow rates, and the need for rapid detector response without peak distortion, all act as severe challenges to *quantitative accuracy and precision*. Detector and column interfacing design must minimize internal detector 'dead volume' and external 'column -detector connecting volume'. Consideration must be given to the effect on practical detection limits of the overall capillary column analytical procedure.

Schill and Freeman (6) note a number of important areas of interaction between the detector and the remainder of the chromatographic system some of which are as follows:

i) *The detector response dependence on the nature and flow rate of the carrier gas.* As already noted, concentration sensitive detectors such as the TCD or the GDB show enhanced sensitivity at low flow rates. For the TCD, differential thermal conductivity between carrier gas and sample determines sensitivity. Helium is to be preferred as a carrier gas but in some areas of work (particularly in Europe) its cost is prohibitive. Hydrogen is usually considered too dangerous although H_2/N_2 mixtures have been used. The FID detector response is somewhat carrier gas dependent but ancillary gases should be pure and impurity levels are critical for optimal signal-to-noise ratios. For many detectors, the nature and purity of the carrier gas is an inherent design parameter. Oxygen and water present in the ECD carrier gas can degrade sensitivity by orders of magnitude. The various other ionization detectors repsonses all depend strongly on carrier gas purity and cleanliness.

ii) *Temperature control of the detector* is critical for most systems to obtain maximum sensitivity. Sample condensation must be prevented and any drift or cycling minimized. Both low sensitivity detectors such as the TCD and GDB and also sensitive detectors, such as the ECD, are highly temperature dependent. Temperature stability to within 1/10°C and the absence of extraneous gradients are critical. Figure 3 illustrates how relative ECD response can vary with detector temperature.

iii) *The electronic procedures* necessary to convert detector parameter response to electrical signal also affect detector performance. The very small currents, nA to pA, obtained from most detectors must be quantitatively transformed to mV level outputs for digital or analog recording and computation.

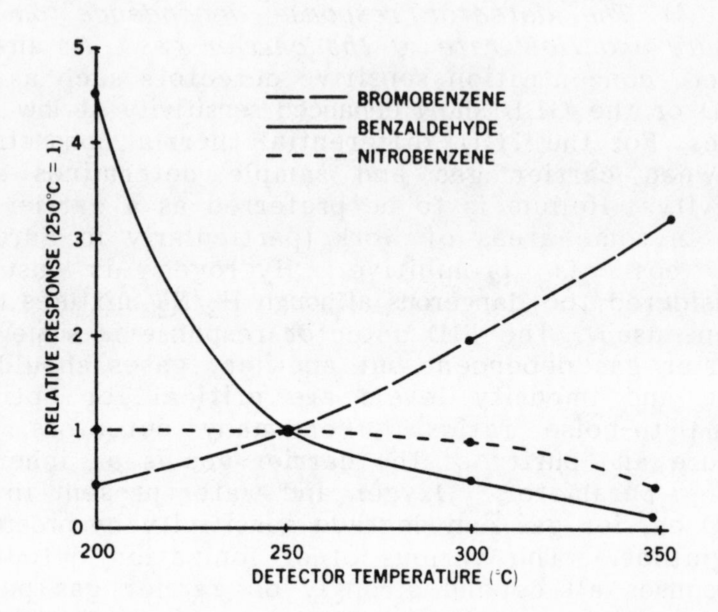

Figure 3. Effect of detector temperature on response of three compounds for ECD.

Improvements in electrometers with low-noise operational amplifiers and transistor feedback resistors in the 10^6 - 10^{11} ohm range make such operations routine but time and temperature stability and cleanliness must always be maintained.

Injection and Sample Introduction

Both sample and component size levels and mode of introduction vary for packed and capillary columns. The total sample volume introduced into the open tubular column is typically 10 nL or less corresponding to a sample mass of less than 10 μg. Peak component levels may thus be 100 ng and lower, as is necessary to preserve column efficiencies for narrow bore capillaries. The effective FID linear response range is thus reduced to no more than 3 orders of magnitude (100 ng - 100 pg) and for the TCD the required capillary peak concentration levels may be below its detection limits. On-line injection -concentration techniques such as adsorption and thermal desorption onto the column may be used to enhance peak signal levels over solvent or interference background and consequently increase effective detector range.

Effect of Dead Volume and Band Broadening on Detection

For packed columns, dead volume arising from detector geometry is minor since the high carrier gas flow rate minimizes the time analyte molecules spend in the sensing volume of the detector. For capillary columns, where volume flowrate may be reduced by a factor of 50 or more, peak widths are much narrower, and severe detector band broadening by longitudinal diffusion can occur especially for peaks eluted at low k' values. Detector sensitivity is consequently significantly affected. Detector residence times can be decreased by one of two methods. The detector volume can be decreased within the limitations of its performance criteria or a carrier gas make-up can be introduced between column and detector. The make-up flow of carrier gas must be optimized to preserve peak efficiency and symmetry.

Multiple Detectors

The utility of multiple detection is being increasingly recognised with the greater use of selective detectors. Figure 4 shows the two major dual detector configurations, series and parallel. Additional multiple detector modes include the conventional dual column 'bleed compensation' system, (to improve baseline drift) and the design in which two fully independent column systems operate simultaneoulsy while sharing the same oven.

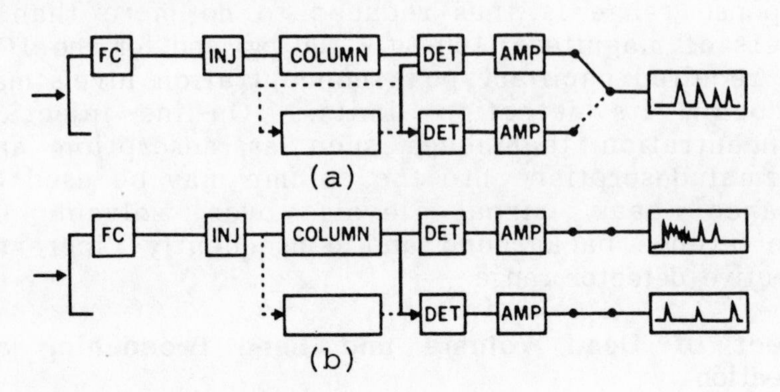

Figure 4. Simplified schematics of practical dual detector and column configrations: a) two detectors in series; b) two detectors in parallel.

Provided the first series detector is non-destructive, a second detector can be directly connected to it or alternatively a second column as in the case of multidimensional analysis. The criteria for peak transfer integrity are most stringent but, nevertheless, dedicated multi-mode instruments are now being introduced. In terms of quantitative accuracy with detector combination, the sequential mode is best where applicable since the same concentration or mass of sample is detected in each. In the more commonly employed parallel or 'multi-channel' mode, either the same or different columns are used to resolve a single injected sample. Direct quantitative comparisons between detectors necessitate individual calibration of both or all channels.

GAS CHROMATOGRAPHY DETECTORS - CHARACTERISTICS AND COMPARISONS

Established, Historical and Developmental Detectors

Since the inception of gas chromatography, as many as 50 viable detection systems have been devised. Of these 10 - 12 have been adopted at various times in commercial instrumentation and a similar number have received recent attention and have appreciable potential for further development . The remaining detectors can be considered instrumental curiosities and are unlikely to be re-examined although it must always be remembered that advancing measurement and electronic technologies may make them worthy of future review. The latter group includes such devices as the titration detector as first used by James and Martin (7), the hydrogen flame temperature detector (8), gas volume detectors based on liquid phase absorption of reactive eluates (9), gas pressure change detectors (10), the piezoelectric sorption detector (11), the dielectric constant detector (12), the semi-conductive thin-film detector (13) and the automatic mass recording detector (14). All of these detectors suffered from a basic lack of sensitivity which was at least worse than the TCD; they would thus be inapplicable for any open tubular column applications.

The Thermal Conductivity Detector (TCD)

The TCD, otherwise known as the hot wire detector (HWD) or the katharometer, was the first generally adopted GC detector since it had earlier achieved wide use for gas analysis and gas composition monitoring. In design for GC, it matured rapidly, its characteristics remaining relatively unchanged as a convenient but relatively insensitive concentration responsive detector with some level of response to all eluates. The detector functions by virtue of resistance change in a heated filament wire or a thermistor (situated in an bridge circuit) when exposed to vapor mixtures of differing thermal conductivity. The response depends upon the resistance change of a filament (or thermistor) with temperature difference across the TC cell which in turn results from a change in thermal conductivity of the surroundings of the filament when in contact with the analyte vapor.

$$\Delta R_r \text{ (resistance change)} \approx R_r(T_1 - T_2)\ \Delta\lambda/\lambda$$

where R_r is the filament resistance,
λ is the thermal conductivity
and T_1-T_2 is the cell temperature differential.

Typically T_1-T_2 is 150-200 °C and R_r is between 25 and 50 ohms dependent upon temperature. For helium carrier gas, the thermal conductivity at 150°C is 4.4×10^{-4} cal/sec.cm.°C.

For an organic analyte peak at the 100 µg level and 10-second peak width, the sample concentration at peak maximum is around 20 µg/mL for a 1.0 mL/sec. carrier gas flow rate. Thermal conductivity at peak maximum would typically diminish by ca. 0.2×10^{-4} cal/sec.cm.°C and the voltage across the filament would typically change by ca. 100 mV. The detector response factor under these circumstances would be 5000 mV.mL/mg (15). Typical values for TCD performance have been proposed by the American Society for Testing and Materials (ASTM) (16). These values are given in Table III and Fig. 5. This publication recommends procedures for TCD evaluation.

General requirements for quantitative TCD performance include good temperature stability, well matched bridge filaments with good electrical insulation and clean carrier gas.

TABLE III

TYPICAL VALUES FOR THERMAL CONDUCTIVITY DETECTOR PERFORMANCE CHARACTERISTICS

Performance Characteristics	Unit	Range of Typical Values
Response factor	mV.mL/mg	5,000 to 15,000
Minimum Detectability	mg/mL	3×10^{-7} to 1×10^{-5}
Linear Range (Upper Limit)	mg/mL	0.2 - 2.0
Dynamic Range	mg/mL	2.1 - 2.4
Noise	μV	5 - 50
Drift	μV/h	20 - 250
Response Time	s	0.5 - 5
Effective Volume	μL	50 - 500
Time Constant	s	0.05 - 0.5

Filament temperature sensitivity is the most critical feature limiting detection, a typical value being the region of 10 mV/°C; a well designed detector can exhibit noise due to temperature variability as low as 1-2 μV.

Three design and operational factors relate to TCD quantitative performance. The use of thermistors shows some advantages over wire filaments, since their response factors are larger due to greater temperature coefficients of resistance change. The MDL, however, remains limited by temperature and pressure control as for filament detection, and

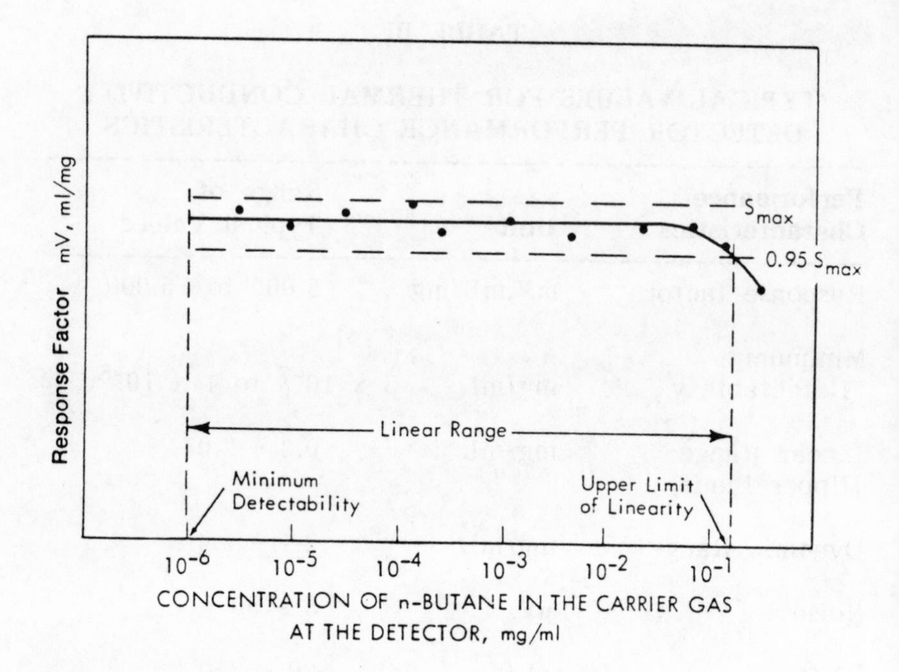

Figure 5. Example for a linearity plot of a thermal conductivity detector

sensitivity diminishes markedly above 50°C. This makes thermistors most suitable for gas analysis applications where condensation in the cell is not a factor. However, thermistors may be miniaturized more readily than filaments and thus have potential in capillary GC detection where fast response and minimal detector cell volume are needed.

Alternative carrier gases to helium may be dictated by operating costs and in fact hydrogen gives somewhat higher sensitivities since its differential thermal conductivity over organic vapors is somewhat greater. However, operational hazards have sometimes necessitated the compromise use of hydrogen/nitrogen mixtures instead of pure hydrogen.

Response factors for various compound classes were investigated in detail by Rosie and Grob (17) and most organics other than small molecules, halogenated compounds and organometallics show factors within ±20% of that of a standard hydrocarbon such as iso-octane.

The Flame Ionization Detector (FID)

The FID is the most widely used GC detector having achieved great popularity in the early 1960s. It is a rugged detector, relatively stable to small changes in operational parameters and with good sensitivity and linearity. It shows similar response factors for all compounds which contain carbon -hydrogen bonds and also has some response to other analytes, while remaining unresponsive to permanent gases or carrier gas impurities. The FID comprises a small hydrogen -air diffusion flame which burns at a capillary jet. A small background current produced by movement of charged species in the flame, induced by the applied voltage impressed across it, is amplified by an electrometer. Introduction of responsive analytes causes a large quantitative increase in flame current although in terms of ionization efficiency only one C-H containing analyte molecule in 10^4 - 10^5 is considered to contribute to the flame current.

Gas flow parameters in the FID are typically in the ratios of: carrier gas/hydrogen/air-1/1/10, although optimization of hydrogen-to-air ratio is often needed since there is an optimum air-hydrogen ratio for maximum sensitivity. Nitrogen is the preferred carrier gas for the use with packed columns with the FID, but helium is also widely used in capillary applications in order to provide enhanced column performance.

Good electrical performance of detector components is critical. Detector voltages up to 300 V are needed to collect ions that provide current levels of picoamps f.s.d. Insulators of 10^{15} -10^{18} ohms resistance are needed at temperatures of 300°C.

The ASTM have proposed standard procedures for testing FID performance (18). In the mass flow detector, response factor is to be measured from the expression

$$R = A_p/m$$

where A_p is the integrated peak area x time and m is the mass of the test substance, preferentially n-butane.

The exponential dilution flask or permeation tube methods are recommended for accurate FID evaluation. Table IV gives typical operational values for the FID detector according to ASTM test conditions.

TABLE IV

Performance Characteristics	Unit	Range of Typical Values
Response factor	Amps.s./g	0.005 - 0.02
Minimum Detectability	g/s	10^{-12} - 10^{-11}
Linear Range	-	10^{6} - 10^{7}
Dynamic Range	-	10^{8} - 10^{9}
Noise	Amps	10^{-14} - 10^{-12}
Drift	Amps/h	10^{-13} - 10^{-12}

Walker, et al. (19) describe quantitative tests carried out on 11 commercial FID detectors, noting operational response factors to range from 0.2 to 5.0 coulomb/g. carbon and noise to range between 1.0 and 25.0 x 10^{-14} amps. By comparison current instrument specifications range closely from 0.012 to 0.020

coulomb/g. carbon and a typical quoted MDQ for pentane is 3×10^{-12} g/s.

FID response factors do vary somewhat among different organic compounds. Heteroatoms and functional groups modify the response predicted according to carbon number. Table V lists some typical contributions to effective carbon number; some functionalities clearly diminish FID response if they are present.

TABLE V

FLAME IONIZATION DETECTOR RESPONSE - CONTRIBUTION TO EFFECTIVE CARBON NUMBER

Atom	Type	Effective Carbon Number Contribution
C	Aliphatic	1.0
C	Aromatic	1.0
C	Olefinic	0.95
C	Acetylenic	1.30
C	Carbonyl	0.0
C	Nitrile	0.3
O	Ether	-1.0
O	Primary Alcohol	-0.6
O	Secondary Alcohol	-0.75
O	Tertiary Alcohol, Ester	-0.25
Cl	2 or More on a Single Aliphatic C	-0.12 each
Cl	On Olefinic C	0.05
N	Amine	similar to O in corresponding alcohol

Modifications to FID Operation

A modified version known as the FIDOH utilizes oxygen instead of air as the combustion gas to give enhanced response to inorganic gases such as H_2S (4 x 10^{-10} g/s), NO (2 x 10^{-11} g/s) and SO_2 (4 x 10^{-10} g/s) (20). Enhanced response to organometallic compounds has been achieved by gas doping with hydrides such as phosphine (21). Similar behavior is noted in the HAFID version (hydrogen atmosphere FID).

The C-H response of the FID may also be used for indirect detection of species which are not directly responsive. Thus oxygen gas is detected at low ppm levels by a two stage conversion process. Oxygen is reacted above 500 °C with carbon to give carbon dioxide which is catalytically reduced to methane for FID detection. Greater than three orders of magnitude of linearity is achievable.

The Alkali Flame Ionization Detector (AFID)

The AFID (sometimes called the thermionic ionization detector (TID) or, from its elemental selectivity, the nitrogen - phosphorus detector (NPD)) relies for its operation on the presence of an alkali metal salt in the vicinity of an FID flame. Alternatively, a heated non-flame orifice, which carries a low hydrogen flow that precludes flame formation, but heats the emitter to a dull red glow at 600-800 °C, can also be employed.

Thermal energy atomizes the alkali metal salt (most frequently rubidium bromide) and alkali metal atoms so formed dissociate into ions and electrons which are subjected to an electrical potential. The presence of halogen, nitrogen or phosphorus containing eluates enhances the ion background current giving a signal proportional to the number of ions present. The full theory of the process is unclear but 'thermionic ionization' is considered the major mode. Other developmental detectors based on this principle are discussed later.

Response factors for the AFID vary for different heteroatoms and are also somewhat functionally dependent. While phosphorus detection has received much attention, nitrogen specific detection is the major application of the AFID. Current specific MDL values for commercial detectors range from 1×10^{-13}-4×10^{-13} g/s for nitrogen in azobenzene and from 1×10^{-14} - 4×10^{-14} g/s for phosphorus in Parathion. Selectivities of N:C (as octadecane) are around 4×10^4 and of P:C are approximately 8×10^4. Linear dynamic ranges in both modes are greater than 10^4. There is no usable selectivity between nitrogen and phosphorus but current detector design shows minimal halogen response.

The Photoionization Detector (PID)

The PID was developed concurrently with the FID but did not gain wide acceptance until design advance enabled the isolation of the UV energy source from the ionization chamber, miniaturization for capillary column interfacing and operation at temperatures up to 300 - 350 °C (22). A range of UV ionization energies are offered extending , from 9.5 to 11.7 eV; these energies lie in the ionization potential range of many molecules, selectivity being energy dependent. The 10.9 eV irradiation provides selectivity among such molecules as methane (12.98 eV ionization potential) and carbon tetrachloride (11.47 eV) which are unionized; formaldehyde (10.87 eV) and methanol (10.85 eV) which are partially ionized, and acetone (9.69 eV) and phenol (8.50 eV) which are sensitively detected. Ionization potential response profiles are shown in Figure 6. The most important practical features of the PID are distinction between aliphatic and aromatic organics and discrimination against small molecules of high ionization potential such as N_2, CO, CO_2, H_2 and He. For maximum selectivity in a particular application, the lamp with energy output just capable of ionizing the target analyte is used.

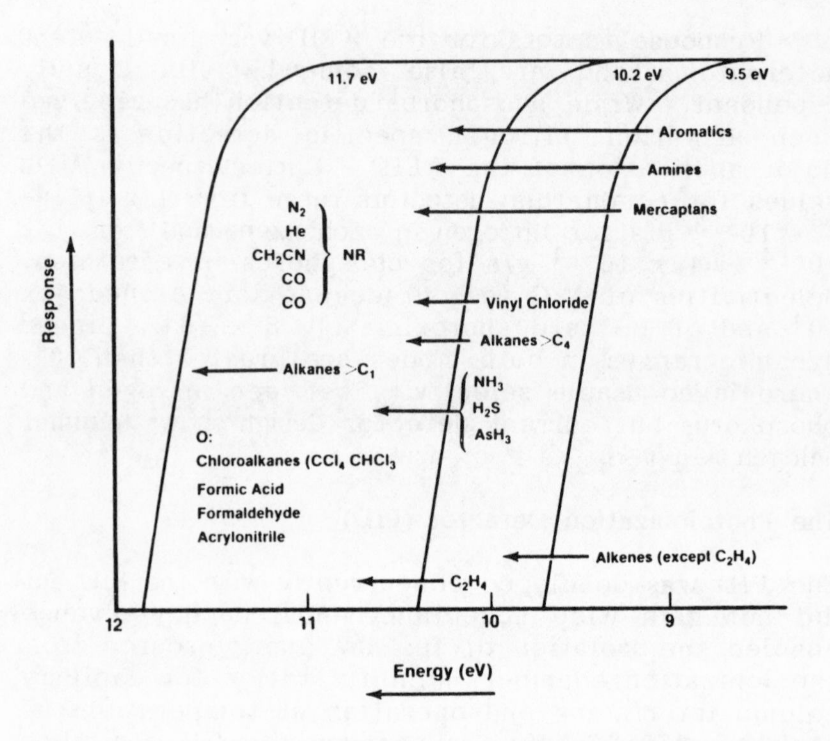

Figure 6. Photoionization detector response for various ultraviolet lamps.

The virtually non-destructive nature of the PID permits serial operation with other detectors. Detector specifications show a linear dynamic range greater than 10^7, a background ionization current of 1.5×10^{-11} amps with a noise level of 4×10^{-14} amps, a response factor of 0.3 coulombs/g of benzene and a MDQ of 2 pg. Detection limits in the range of 10 -200 pg are obtainable for inorganic molecules such as phosphine, ammonia, arsine and tetraethyl lead. The acceptance of the PID is evident as it is the chosen detector for the US Environmental Protection Agency standard GC method for purgeable aromatics in drinking water.

The Electron Capture Detector (ECD)

Among the ionization - based GC detectors, the greatest sensitivity and molecular selectivity has been obtained with the ECD. It is now considered a routine detector, since both the design and operation have been simplified to reduce problems evident in earlier models . The ECD was invented by Lovelock and evolved from the argon ionization detector (23). Lovelock imposed a 1000 volt potential across a pair of electrodes in an argon stream. Particles (electrons) from a radioactive source generate metastable argon (11.6 eV energy) which ionizes many substances by a collisional process: $Ar^* + M \rightarrow Ar + M^+ + e^-$.

In the ECD detector the argon was replaced by a 5% methane in argon (to quench the production of metastable atoms) or nitrogen which produced no metastable atoms at all. The major advantage of this detector is that primary β electrons generate up to 10^4 secondary electrons, and that electronegative species containing halogen atoms or other functional groups capture electrons: $MX + e^- \rightarrow MX^-$. This 'electron capture' phenomenon reduces the standing current and provides the GC detection signal.

Various primary ECD electron sources have been used, for example tritium or ^{63}Ni which are weak beta emitters, and some non-radioactive emitters have also shown promise. The preferred carrier gas is a 90:10, argon:methane mixture which exhibits rapid electron diffusion but, as already stated, quenches the production of metastable atoms. Response factors in ECD depend on a large number of parameters. Detector output current has been expressed as:

$$I = I_s/\Phi(1-e^{-\Phi})$$

where I_s is the current produced by the radioactive source and Φ is a proportionality constant.

ϕ depends upon the rate constant of electron capture, the rate constant of reassociation of electrons and positive ions, sample concentration and output frequency. Rate constants of the many competitive processes and other factors are also included in equations derived by more rigorous theoretical treatment.

Limited linear dynamic range has been a negative feature of ECD operation, frequently being as small as two orders of magnitude. Operation in a constant current mode can provide automatic control of anode pulse frequency to increase linearity to 10^4. A typical MDL value is 10^{-13} g of Lindane per mL of nitrogen carrier gas at a detector temperature of 250°C. Highly electron capturing species have exhibited spectacular sensitivities; sulfur hexafluoride to 10^{-14} g/s MDL and N,N dipenta-fluorobenzoylpentafluoroaniline at 9 x 10^{-17} g MDQ (24). Based on a relative response factor of 1.0 for chlorobenzene, representative values range from n-hexane (0.01), naphthalene (0.1), acetophenone (10), benzyl chloride (ca. 100), chloroform (ca. 1000) and carbon tetrachloride (ca. 10,000).

The Mass Spectrometer Detector (MSD)

The MSD is the most widely used detector for GC which involves an independent analytical instrument of higher inherent complexity than the chromatograph itself in a tandem - interfaced configuration. The mass spectrometer provides a true tandem 'second separation mode' to the analysis since it operates to separate, quantitate and identify analyte ions which derive from GC eluates.

In some respects the gas chromatograph and the mass spectrometer have high compatability. Each functions at similar analyte concentration levels. The principal interface problems involve quantitative peak transfer to the mass spectrometer and the elimination of most of the carrier gas molecules in order to retain a suitable vacuum in the MS ionization source. The latter is of importance in any interface design. Various 'enrichment' interfaces have been adopted but at present two systems dominate. These are the 'jet separator' and the direct capillary column interface. The latter employs high efficiency turbomolecular vacuum pumps to obviate sample loss resulting from the reduction of column exit pressure to meet source requirements. The jet separator, although less quantitative, can give up to 50% effective sample transfer.

Mass spectral ionization modes utilized in GC-MS are primarily 'electron impact (EI) and 'chemical ionization' (CI). With 70 eV EI, a full mass spectrum may be obtained from 1 -10 ng of a compound in a GC eluant. If the mass spectrometer is utilized in the 'selected (single) ion monitoring mode (SIM)', wherein it remains focussed on a single abundant ion throughout the GC profile, 10-100 pg GC peaks can be detected. The third way in which the GC-MS system is typically used is the 'total ion monitoring' (TIM) mode in which the sum of the ions at all masses detected at a given time in the chromatogram is displayed. The TIM GC-MS mode is considered to act as a 'universal detector' while the SIM mode acts as a selective detector. These alternative modes of operation, coupled with mass spectral identification by EI or CI, emphasize the versatility of the GC/MS system.

Operational specifications for quantitative GC-MS detection depend on various aspects of the system. These include interface efficiency with regard to peak transfer and profile preservation, ionization mode and efficiency, analyzer type, i.e. magnetic sector, multiple sector, single or multiple quadrupole, and finally the computer data management system. A recent feature of GC-MS development has been the introduction of integrated mass spectrometers designed specifically as GC detectors only. The 'ion trap' and mass selective detectors are examples of this trend to bring GC-MS more directly into the providence of the chromatographer.

The Infrared Spectrometer Detector

The spectral analysis most widely used in organic chemistry has been infrared spectrophotmetry. The utility of IR as a GC detection mode is clearly for functional group identification as an aid to the structural elucidation of an eluate. Since analytical vapor phase IR (VPIR) has received much less emphasis than condensed phase methods, early GC procedures involved sample trapping from the column. As a consequence poor quantitative sample transfer and modest sensitivity of the dispersive spectrometers, that were then used, significantly reduced the effectiveness of this approach. The advantage of on-line vapor-phase analysis is clear and rapid-scan (6-30 seconds) dispersive spectrometers allowed usable VPIR spectra to be obtained from packed column GC eluates at sample levels of a few μg, over a wavelength range of 2.5 to 15 μm. Typical functional group absorptivities, however, preclude the use of dispersive spectrophotometers for capillary column application.

The inherent sensitivity disadvantages of GC-IR as compared with GC-MS have been *partially* overcome by the development of Fourier transform methods (FTIR). Computer based interferometer procedures provide up to three orders of magnitude sensitivity enhancement to the interfaced technique. Analytes which contain strongly IR absorbing functionalities may now be estimated quantitatively in GC eluates down to the low ng range. Interfaces have been developed, often utilizing gold coated multiple reflectance light pipe cells of sufficiently low volume to minimize band spreading for capillary column peaks.

GC-FTIR is used more for *qualitative* analysis, but data systems allow excellent quantitative monitoring of individual or selected groups having unique absorbing wavelengths. Computer data handling allows signal averaging, background subtraction and search and comparison programs utilizing VPIR spectral libraries. Some highly sophisticated analytical approaches have combined GC-FTIR and GC-MS, the non-destructive nature of the former allowing sequential or parallel detector interfacing (25).

The Flame Photometric Detector (FPD)

The FPD is an element selective, solute property, mass flow sensitive detector that has been applied almost exclusively to the specific monitoring of sulfur and phosphorus containing compounds. It was the first widely employed spectral detector for GC. Eluting compounds are combusted in a relatively cool hydrogen-rich flame to form chemiluminescent molecular species. In some designs two flames in series combust samples and provide the energy for molecular emission. Broad band optical emission is spectrally isolated by interference filters, wavelengths centered at 394 nm for sulfur (the S_2 species is the primary molecular emitter) and 526 nm for phosphorus (the HPO species being the major emitter). Photomultiplier tube detectors are thermally isolated from the flame to enhance signal to noise.

Response for phosphorus is linear with concentration due to the formation of one HPO per phosphorus atom; selectivity over hydrocarbons is as great as 10^5. Selectivity of phosphorus over sulfur, however, is low, in the range 5 - 10. The inverse selectivity of sulfur over phosphorus, however, is as high as 10^4 and thus a good distinction can be made between the two elements, if necessary, by making measurements with both interference filters.

The response of sulfur is determined as a square root relationship since the flame reaction converts sulfur atoms to S_2 species. A linearizing circuit is usually incorporated in the detector electronics to accomodate this. The differing response factors also provide a method to discriminate between the elements when both are present in an analyte compound. Selectivity of S/hydrocarbon is typically 10^4. MDQ values for commercial detectors range for 10^{-10} to 2 x 10^{-11} g S/s and from 10^{-11} to 10^{-12} g P/s, making the detector somewhat more sensitive than the FID for sulfur and phosphorus analysis. Linear dynamic ranges are variously quoted from 10^2 to 10^4 for sulfur and from 10^3 to 10^7 for phosphorus. The molecular bond energies of organically bound sulfur and phosphorus are sufficiently low as to minimize any functionality dependence of element response.

The Electrolytic (Hall) Detector (HECD)

Among electrometric (solution phase) GC detectors, the electrolytic conductivity detector which monitors ionic species in water has been the most extensively developed. The good detection limits and selectivities otainable by this detecting system have overcome the inherent complexity of the detection process. The analyte is initially converted by oxidative or reductive pyrolysis in a high temperature micro-reactor into ionic species which are transferred to a stream of deionized solvent for electrolytic detection. Molecular chlorine is converted to hydrogen chloride, sulfur to sulfur dioxide and nitrogen to ammonia in three separate reaction modes. Detection specificity results from the choice of solvents, reaction conditions employed and the use of scrubbers appropriate for interference removal in any particular mode of detection (26).

The present specification of the HECD are comparable with any other element selective detectors for nitrogen, sulfur and total halogen. MDL levels of nitrogen are 1-2 x 10^{-12} g N/s, for sulfur 5 -10 x 10^{-13} g S/s and for halogen (as chlorine) 1-2 x 10^{-13} g Cl/s. Linearity which is dependent upon quantitative conversion of the analyte element is 10^4 for nitrogen and sulfur and 10^6 for chlorine. Selectivities over hydrocarbons are 10^5 for sulfur and 10^6 for nitrogen and chlorine. Inter-element selectivities are extremely high since reaction modes are distinct, the exception is that halogens are detected as chlorine and thus, inter-halogen response factors require further study. Miniaturization of the system permits full compatibility with capillary column separations.

SPECIAL DETECTION SYSTEMS

In this section those GC detectors are described which may be considered more specialized, innovative or are open to further development to realize potentially wider use. All have been made commercially available at one time or the other

The Gas Density Balance Detector (GDBD)

This universal, bulk property, concentration responsive detector was conceived by the Noble Laureate A.J.P. Martin and has received recent attention in a dual column mode for direct density -differential based molecular weight determination (27). The sample densities of the reference gas and GC eluate are measured by the differential flow induced by their density difference, in a Wheatstone bridge system of capillary tubes. Response factors depend on molecular weight and density of the carrier gas used in a matched two - column configuration; carbon dioxide has been favored for higher molecular weight analytes and Freon™-C_2ClF_5 for lower molecular weight compounds. Calibration against standard molecular weight compounds has permitted the determination of the molecular weight of an unknown sample to better than one mass unit. There remains some functionality dependence and very accurate column matching is needed. The major operational drawback has been the limited detection sensitivity which parallels that of the TCD and has thus made capillary column application impractical.

Modified Thermionic Ionization Detectors (TID)

The success of the widely used NPD detector for nitrogen and phosphorus has prompted much detailed study of surface ionization mechanism and systematic variation of key parameters, notably the work function of the thermionic surface. As a result, many other modes of detector response have been developed (28). While various forms of alkali metal themionic sources have been used, the most versatile comprise of multiple layers of cylindrically shaped ceramic coating with a non-corrosive electrically conducting sub-layer of nickel, ceramic covering a loop of nichrome wire and a surface layer comprised of alkali and/or other additives in a ceramic matrix. Three different chemical compositions have been shown to be particularly useful, their appliciation also involving different combustion gas conditions.

The simplest system involves a low work-function thermionic source, which incorporates a high cesium salt concentration and is operated in a nitrogen environment. The TID surface is heated to 400 -600 °C and the ionization process involves direct transfer of negative charge to the sample molecule. High sensitivity and specificity to nitro functional groups and certain other electronegative compounds have been noted. Selectivity is quoted as high as

10^8 for methyl parathion against n-pentadecane. MDQ values from 0.1 to 1.0 pg/s are obtainable for dinitrotoluene and linearity is greater than 10^3.

A therminoic source incorporating a low concentrations of cesium and strontium gives specific performance characteristics in two analytically useful modes. When operated in an oxygen -containing gas environment, specific response to halogenated compounds is enhanced and that to nitro compounds is diminished. Typical selectivity of halogen over hydrocarbon is 10^4 and MDQ values are 0.1 - 1.0 ng of halogen. The linearity covers a range of 10^3 and extends to higher solute concentrations than those at which the ECD and HECD are saturated. The same thermionic source exhibits good nitrogen and phosphorus selective response in a hydrogen/air environment at 600-800°C. Selectivity at a hydrogen flow of 3 mL/min is 2×10^4 for N/C as azobenzene/n-heptadecane, and 4×10^4 for P/C as malathion/n-heptadecane; there is little N/P selectivity, however. The MDL for azobenzene is 10^{-13} gN/s and linearity for nitrogen is 10^5. An advantage is the high possible operating temperature of 420°C. If a higher hydrogen flow is used in conjunction with a nickel-containing ceramic, a self-sustaining flame is produced and the operating mode is designated as catalytic flame ionization detection (CFID). Most organic compounds and heteroatom compounds show MDQ levels in the 10-100 pg range. It seems probable that further fruitful detectors may be developed in this class as other catalyst systems are investigated.

Inert Gas (Helium and Argon) and Universal Plasma Ionization Detectors

These detectors operate by ionization of analyte molecules by excited metastable rare gas atoms produced by interaction with acelerated electrons derived from β-radiation from an appropriate radioactive source, typically tritium. Since the energy of helium metastable atoms (19.8 eV) is higher than those of argon (11.6 eV), the former can ionize more analyte species, in fact virtually all molecules passing through the detector. Ions formed are subjected to a high electric field (200-2000 V/cm^2 of electrode surface) and current changes are measured by an appropriate current amplifier. Although the argon detector was devised earlier, the helium detector has gained greater acceptance primarily for permanent gas analysis by gas solid chromatography. The emphasis on gas solid chromatography is because extremely pure helium carrier gas is required to minimize metastable losses and stationary phase bleed cannot be tolerated. MDL levels for nitrogen below 10 ppb and for carbon dioxide and methane below 5 ppb are reported with linearity of 10^4. The MDL for some analytes is as low as 4×10^{-14} g/s. Relative response may be modified by varying the cell voltage. Changes in carrier gas and other operation parameters can transform the helium ionization detector (HID) into a 'universal' plasma ionization detector (UPID); argon carrier gas has proven useful for detection of hydrogen, oxygen and nitrogen gases at the 100 ppb level.

Atomic Spectroscopic Detectors

The most widely applicable element specific detectors for GC are those which employ atomic spectroscopy with wavelength specific spectral intensity monitoring. Three types of atomic spectroscopy have been interfaced successfully as GC detectors, atomic absorption, flame emission and plasma emission. The first two have seen limited application, but inert gas plasma emission methods show considerable analytical potential despite the substantial spectroscopic instrumentation needed which, unfortunately, makes the complexity of the apparatus similar to GC-MS or GC-IR.

The Flame Emission Detector (FED)

A high temperature oxygen flame has sufficient energy to atomize molecular structures and excite a number of elements to emit their characteristic spectral wavelengths. Juvet and Durbin (29) determined a number of metals, including iron and chromium in volatile metal chelates, the degree of element specificity being dependent upon the resolution of the monochromator used. Selectivity over carbon was 10^3-10^4 and MDQ levels were 10^{-6}-10^{-9} g of metal. The combined operation of a FED as a FID has been described.

Atomic Absorption Detector (AAD)

Flame AA has been quite widely applied to the problem of lead specific detection. Measurement at 283.3 nm gave a 20 ng detection limit for lead in alkyl lead compounds and a linearity of 10^3. Detection limits for solid samples were between 0.01 and 0.025 µg/g (30). Graphite furnace atomization at 1500 °C has also been interfaced with GC for detection purposes as has a tantalum furnace at 2500°C. Alkyl lead concentrations at around 70 ng/m^3 in air have been monitored by a trapping procedure. Other metals successfully determined by GC-AAD have been arsenic, selenium, chromium and mercury. Cold vapor AA analysis for the latter has been the preferred method. The inherent limitation of the AAD is the measurement of only one single target element in any one chromatogram.

Atomic Emission Detector (AED)

In contrast to AAS, atomic emission spectroscopy has the advantage of being a multi-element technique with a wide dynamic range of measurement . The advent of various accessible plasma sources, particularly in combination with high resolution monochromators to minimize spectral interferences, has produced a resurgence of analytical applications, not least in chromatographic detection. The three major sources used in GC detection have been the microwave induced and sustained helium plasma (at reduced or atmospheric pressure) (MIP), the DC argon plasma (DCP) and the inductively coupled argon plasma (ICP). The major advantages of GC-plasma AES are: a) the ability to perform speciation, either prior to or within the chromatographic column for many metals and non-metals, directly or through derivatization; b) the ability to tolerate non-ideal chromatographic elution: the specificity of plasma emission enables the chromatographer to tolerate incomplete resolution, common for complex matrixes; c) the frequent high sensitivity achieved; d) the multi-element capacity; e) the compatibility with GC systems using simple interfaces.

The Microwave Induced Plasma (MIP)

Early atmospheric pressure GC-MIP systems used an argon plasma induced and sustained by a 2450 MHz microwave source, but later, helium plasmas at reduced pressure were more favored. McLean et. al (31) described a comprehensive multi-element system operated at 0.25 torr; detection limits ranged from 10^{-7} to 10^{-12} g/s for elements such as the halogens, phosphorus and sulfur but selectivities against carbon were poor, in the range of 10-100. Practical limits of 0.03 ng/s for hydrogen and 3.0 ng/s for oxygen were noteworthy.

A major advance in GC-MIP was made by utilizing a TM_{010} cylindrical resonance cavity which permits, at atmospheric pressure, helium plasma to be maintained at the same low power as for reduced pressure plasmas. The cavity configuration is ideally suited to capillary column interfacing. A comprehensive listing of detection limits and selectivities is given in Table VI (32). Linear dynamic ranges for most elements are in the range of

TABLE VI

DETECTION LIMITS AND SELECTIVITIES FOR CAPILLARY GC MIP DETECTION

Element and Emission Wavelength nm	MDL pg/s	Selectivity vs C.	Element and Emission Wavelength nm	MDL pg/s	Selectivity vs. C.
H 486.1	16	74	Al I 396.2	5.0	3.90×10^3
H 656.3	7.5	160	C I 247.9	2.7	1.00
D I 656.1	7.4	194	Si I 251.6	9.3	1.58×10^3
V II 268.8	10	5.69×10^4	Ge I 265.1	1.3	7.57×10^4
Nb II 288.3	69	3.21×10^4	Sn I 284.0	1.6	3.58×10^5
Cr II 267.7	7.5	1.08×10^5	Pb I 283.3	0.17	2.46×10^5
Mo II 281.6	5.5	2.42×10^4	Pb I 405.8	2.3	2.00×10^5
W II 255.5	51	5.45×10^3	P I 253.6	3.3	1.06×10^4
Mn II 257.6	1.6	1.11×10^5	As I 228.8	6.5	4.70×10^4
Fe II 259.9	0.28	2.80×10^5	S II 545.4	52	4.59×10^3*
Ru II 240.3	7.8	1.34×10^5	Se I 204.0	5.3	1.09×10^4
Os II 225.6	6.3	5.00×10^4	F I 685.6	180	1.14×10^4*
Co I 240.7	6.2	1.82×10^5	Cl II 479.5	86	1.49×10^3*
Ni II 231.6	2.6	6.47×10^3	Br II 470.5	67	1.06×10^3*
Hg I 253.7	0.60	7.69×10^4	Br II 478.6	34	599
B I 249.8	3.6	9.25×10^3	I I 206.2	21	5.01×10^3

*With refractor plate background correction

2×10^2-10^3. A typical chromatogram showing carbon and lead specific detection is shown in Figure 7.

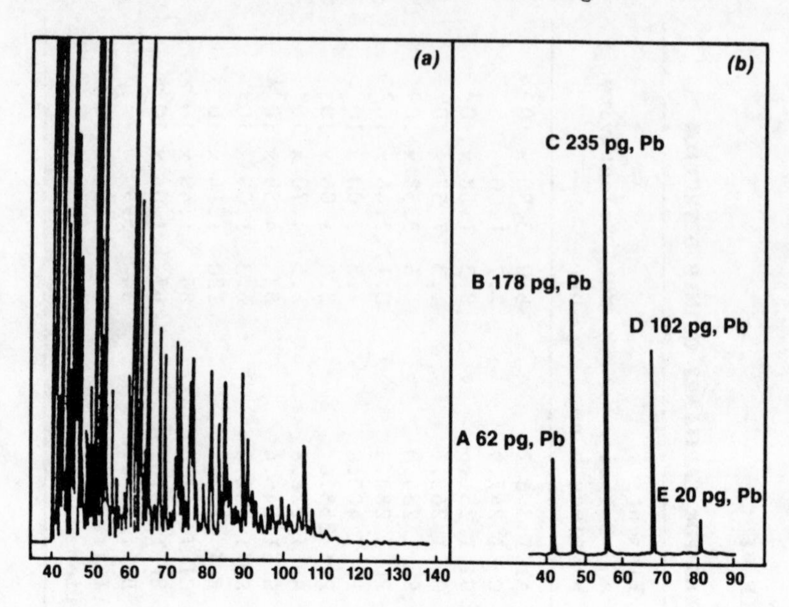

Figure 7. Microwave induced atomic plasma emission (MIP) GC detection. Gasoline (Super Shell leaded), 0.10 µL, split 100:1 to a 12.5 m SP-2100 fused silica capillary column. a) Carbon specific detection at 247.9 nm. b) Lead specific detection at 283.3 nm. Peaks A. Me_4Pb; B. Me_3EtPb; C. Me_2Et_2Pb; D. $MeEt_3Pb$; E. Et_4Pb.

Selectivity is primarily a measure of spectral resolution and depends on the wavelength range of emission which may be measured by a particular spectrometer. Values are typically very high for metals but poorer for some non-metals, however, extension of the determination into the vacuum ultraviolet or the near infrared regions shows promise.

The DC Argon Plasma (DCP)

The wide use of the atmospheric pressure DC argon plasma as a spectral source encouraged its application as a GC detector. Detection limits for most metals, boron and silicon are within an order of magnitude of those of the MIP and selectivities over carbon are typically 10^6 - 10^8. Sensitivity for non-metals however is low, so detector applications are more restricted.

The Inductively Coupled Argon Plasma (ICP)

In comparison with the MIP and DCP, the ICP has been little utilized as a GC detector, but in view of the great popularity of the ICP technique, it may be developed in the future. An elemental analysis procedure for empirical formula determination has been reported (33), with detection limits in the low ng range for metals and some non-metals.

The Ultraviolet Absorption Detector

Ultraviolet spectrophotometers have been used for vapor phase detection after chromatographic eluent condensation, but measurements in the 200-350 nm range generally lacked sensitivity for most organic compounds which have only modest molar absorptivities in that region. Multiple-ring aromatic systems have given the greatest sensitivity but typical detection limits were 100 ng - 1 µg, precluding use in capillary column applications. Little effort has been expended to obtain full on-the-fly vapor phase UV spectra for GC eluates since their information content for identification purposes is modest.

If measurement is extended to the far UV (FUV) region however, down to 120 nm two major changes are observed. The detector is now virtually universal responding to all but rare gases, response factors varying by no more than a factor of 3-4 for any organics. Typical detection limits are in the 4 - 10 ng range for E = 10^4, and linearity is around 10^4. The low dead volume of 40 µl makes the detector suitable for capillary column use and its non-destructive mode facilitates sequential detector configurations.

The Ultrasonic Detector (USD)

In this detector sound waves are propagated between a pair of gold-plated ceramic transducers. Two such cells are set up to produce a phase difference which is transformed to a signal output. When a GC peak is introduced, the change in gas composition produces a phase angle change which is related to the analyte molecular weight, the carrier gas molecular weight, the gas constant, specific heat and temperature. The latter must be held constant to 10^{-4}°C at lower detection limits. The major utility of the detector is in trace permanent gas analysis by GSC since it does not function well at elevated temperatures. A linear dynamic range of 10^6 is obtainable and MDQ values for hydrogen at 2 x 10^{-9} g and ethane at 1 x 10^{-9} g are noted with helium carrier gas. The closed leak free detector system also suggests utility for corrosive and flammable gases and the absolute nature of the detector has been used to calculate absolute weights of unknown analytes without calibration curves (34).

Chemiluminescent Detectors

This group of detectors is achieving increasing prominence. Although the FPD functions by generation of chemiluminescent S_2 and HPO species, the first commercial detector to use an on-flame generated specific chemiluminescent reaction was the 'thermal energy analyzer' (TEA), a specific detector for the measurement of N-nitroso compounds. The sample is pyrolyzed to form nitrosyl radicals which are reacted with ozone to form excited nitrogen dioxide which on chemiluminescent relaxation emits near IR radiation. Linearity is 10^6 and detection limits for N-nitroso compounds are 10^{-9} and less, the high lability of the nitroso group provides similar response factors for different analytes.

A recent development of the detector is the redox chemiluminescence detector (RCD). If nitrogen dioxide is added continuously to a GC column effluent, eluting compounds can be rapidly catalytically dehydrogenated or oxidized forming pulses of nitric oxide which are sensed downstream by the chemiluminescent reaction with ozone (35). Reaction selecitivity is achieved by control of the catalyst temperature and the contact time. Reproducibilitly of 1.2% for carbon monoxide detection by closed loop injection required temperature control at 400 ±2°C. A group of analytes which do not respond in this detector include water, nitrogen, carbon dioxide, alkanes and chloroalkanes. A typical detection limit for 2,6-dimethylphenol of 200 pg is quoted and relative response factors range by comparison from 0.2 to 2.5. A linearity range from 200 pg to 100 ng is typical.

Another chemiluminescent detector that has been evaluated for the analysis of nitrated polycyclic aromatic hydrocarbons involves their pyrolysis at 1000°C to generate nitrosyl radicals with subsequent chemiluminescent reaction with ozone. Detection limits of 50 pg are noted.

These detectors combine chemical reaction principles with sensitive specific detection and typify the future direction of compound and functionality selective GC detection.

SYNPOSIS

A separation process must be accompanied by

quantitative interpretation, qualitative assessment is inadequate, consequently an appropriate dedicated detection system is essential. GC provoked the development of the first detectors that sensed the presence of organic vapors in gas streams at very low concentrations. The properties sensed by the different detectors covered a wide range from thermal conductivity to various ionization techniques. *Appropriate detectors* must have *high sensitivity, wide linear dynamic range and not impair the chromatographic separation*. There are various methods of classification, some based on the detection process, others on the method of measurement; about *fifteen different detectors are commercially available at this time*. There are two basic types of detector, *mass sensitive detectors* and *concentration sensitive detectors*. Universal detectors detect all types of substances (the TCD), but exhibit limited sensitivity; selective and specific types of detector detect only special types of substances (the ECD), but at very high sensitivity. Ionization detectors promote the ionization of the solute by some suitable means and monitor the resulting ionic current. Bulk property detectors measure some physical property of the effluent such as thermal conductivity and provide a proportional voltage output. Spectroscopic detectors measure some optical property of the solute such as infrared absorption. Similarly, electrochemical and reaction detectors monitor electrical and chemical reaction characteristics of the solute. The *important specifications of a GC detector* are the same as those for an LC detector, *noise level, minimum detectable quantity (MDQ), response, linear dynamic range, detector volume and dispersion*. Universal standardization is not possible as different detectors give quite different responses to the same compound. Some detectors are appropriate for packed columns, others for capillary columns, depending on their speed of response and dispersive characteristics. *Ideally detectors should be insensitive to flow rate changes, temperature changes and injection procedures.* The use of multiple detectors can enhance solute identification and in multidimensional analysis, resolution. The two most commonly used detectors are the TCD and the FID. The TCD has a very limited sensitivity and linear dynamic range and is used mainly in the analysis of permanent gases. The FID is the work horse of GC; it has a very good sensitivity, a wide linear dynamic range and with very few exceptions detects all compounds containing carbon. The AFID incorporates a bead of alkali salt closely associated with the flame of a FID (usually rubidium chloride) which renders the detector highly sensitive to substances containing phosphorus, nitrogen and/or halogen. It has a reasonable liner dynamic range. *The PID employs low wavelength UV light to ionize eluted solute vapor,* the ions being collected and measured by an appropriately placed electrode system and collection voltage. *By choosing the UV wavelength the detector can be made solute selective,* with high sensitivity and adequate linearity. The ECD is also a highly selective detector with very high sensitivity, limited linearity and highly specific to halogen compounds. *The MS detector,* in the total ion monitoring mode, is *probably the nearest to the universal detector.* The mass spectrometer and the gas chromatograph are highly compatible and work well in tandem. The combination is probably the most powerful dual technique for separation and structural identification of the individual components of a mixture. The *IR spectrometer detector, even the FTIR system,* is *much less sensitive than* the GC/MS and less informative but on occasion *can provide valuable support to the GC/MS.* The FPD is chemiluminescent detector specific for sulphur and phosphorous compounds. It is sensitive and reasonably linear. The HECD or Hall detector is highly sensitive to nitrogen, sulphur and halogens, but requires chemical reaction prior to detection and consequently requires carefully designed apparatus to prevent instrument dispersion. Some special detectors worthy to note are the Gas Density Balance and the Inert Gas Detectors, e.g. the Argon Ionization Detector. *The GDB* has very limited sensitivity *but can measure the molecular weight of eluted solutes to an accuracy of one mass unit*. The inert gas detectors have high sensitivity, restricted linear dynamic range and tend to be universal in response. The Helium Ionization Detector will detect virtually all compounds but is very sensitive to column bleed and is thus largely and almost solely used in Gas Solid Chromatography. *Atomic spectroscopic detectors can provide element specificity with high sensitivity* and reasonable linearity, particularly for the metal elements. The atomic absorption spectrometer has a sensitivity between 10-25 pg/gm, unfortunately, can only monitor single elements at a time. *The atomic emission spectrometer*, in contrast, *provides* similar sensitivity, but also *multi-element capability*. The microwave induced plasma detector also provides multi-element capability as does the DCP and ICP. The advantages of the element specific detector lie in their use with poorly resolved mixtures where the peak containing the element is singled out from a complex of many eluted peaks. The UV absorption detector has adequate sensitivity but only for aromatic type materials. The ultrasonic detector has a wide linear dynamic range, relatively poor sensitivity and is used only for permanent gases. The chemiluminecent detectors are again highly specific, reasonably sensitive and have the advantages of dealing with solutes that are very poorly resolved but clearly apparent due to the specificity of the detector response.

References

1. J.A. Perry, "Introduction to Analytical Gas Chromatography", Marcel Dekker, New York, 1981, p. 144.

2. M.J. O'Brien, in "Modern Practice of Gas Chromatography", R.L. Grob (Ed.), Wiley, New York, 2nd ed., 1985, p.222,

3. H. Oster and F. Oppermann, Chromatograhia, 2 (1969) 251.

4. F.J. Yang and S.P. Cram, J. High Resolut. Chromatogr. and Chromatogr. Commun., 2 (1979) 487.

5. D.J. David, "Gas Chromatography Detectors", Wiley, New York, 1974.

6. R. Schill and R.R. Freeman, in "Modern Practice of Gas Chromatography", R.L. Grob (Ed.), Wiley, New York, 2nd ed., 1985, p. 337.

7. A.T. James and A.J.P. Martin, Analyst, 77 (1952) 915.

8. R.P.W. Scott, in "Vapor Phase Chromatography" D.H. Desty (Ed.), Academic Press, New York, 1957.

9. J. Janak, Microchim. Acta, 1038 (1956).

10. F. Van de Craats, Anal. Chim. Acta, 14 (1956) 136.

11. W.H. King, Jr., Anal. Chem., 36 (1964) 1735.

12. G. Johansson, Anal. Chem., 34 (1962) 914.

13. T. Seiyama, K. Fujiishi, M. Nagatani and A. Kato, Chem. Abstr., 60 (1964) 1083d.

14. S.C. Bevan, T.A. Gough and S. Thorburn, J. Chromatogr., 43 (1969) 192.

15. M. Dimbat, P.E. Porter and F.H. Stross, Anal. Chem., 28 (1956) 290.

16. "Standard Recommended Practice for Testing Thermal Conductivity Detectors used in Gas Chromatography", ASTM E 516-74, Philadelphia PA, 1974.

17. D.M. Rosie and R.L. Grob, Anal. Chem., 29 (1957) 1263.

18. "Standard Recommended Practice for Testing Flame Ionization Detectors used in Gas Chromatography", ASTM 574-77, Philadelphia PA, 1977.

19. J.Q. Walker, S.F. Spencer and S.M. Sonchik, J. Chromatogr. Sci., 23 (1985) 555.

20. P. Russev, M. Kunova and V. Georev, J. Chromatogr., 178 (1979) 364.

21. M.D. Dupuis and H.H. Hill, Jr., J. Chromatogr., 195 (1980) 211.

22. J. N. Driscoll, J. Chromatogr. Sci., 23 (1985) 488.

23. J.E. Lovelock, J. Chromatogr., 1 (1958) 35.

24. R.J. Gordon, J. Szita and E.J. Faeder, Anal. Chem., 54 (1982) 478.

25. R.W. Crawford, T. Hirschfeld, R.H. Sanborn and C.M. Wong, Anal. Chem., 54 (1982) 817.

26. R.C. Hall, J. Chromatogr. Sci., 12 (1974) 152.

27. E. Kiran and J.K. Gillham, Anal. Chem., 47 (1975) 983.

28. P.L. Patterson, J. Chromatogr. Sci., 24 (1986) 41.

29. R.S. Juvet and R.P. Durbin, Anal. Chem., 38 (1966) 565.

30. Y.K. Chau, P.T.S. Wong and P.D. Goulden, Anal. Chim. Acta, 85 (1976) 421.

31. W.R. McLean, D.L. Stanton and G.E. Penketh, Analyst, 98 (1973) 432.

32. S.A. Estes, P.C. Uden and R.M. Barnes, Anal. Chem., 53 (1981) 1829.

33. D.L. Windsor and M.B. Denton, J. Chromatogr. Sci., 17 (1979) 492.

34. K.J. Skogerboe and E.S. Yeung, Anal. Chem., 56 (1984) 2684.

35. M.J. Bollinger, D.W. Fahey, F.C. Fehsenfeld and R.E. Sievers, Anal. Chem., 55 (1983) 1980.

Quantitative Analysis using
Chromatographic Techniques
Edited by Elena Katz

Chapter 5

QUANTITATIVE ANALYSIS BY GAS CHROMATOGRAPHY

Charles E. Reece and Raymond P. W. Scott

Quantitative analysis has been an important aspect of gas chromatography separations since the inception of the technique in the early 1950's. The general procedure for carrying out the quantitative estimation of each sample component starts with adequate resolution. However, for *accurate* quantitative analysis greater resolution will be needed than for qualitative analysis to allow the peak areas or peak heights to be measured with sufficient accuracy. The degree of resolution that is required will, of course, depend on the particular type of peak measurement being used. There are, however, four key factors in obtaining accurate quantitative analyses and they are, careful sample preparation, precise sample injection, adequate resolution of the mixture and the use of a detector having a known and reliable linear response. If these four factors are attended to, the chromatographic measurements will be both accurate and precise and there are a plethora of data processing methods, both manual and electronic that can take this data and provide an accurate analysis.

SAMPLE PREPARATION

In any quantitative analysis the aliquot of sample taken to be analyzed must be representative of the whole and be accurately and precisely selected as with any other analytical technique. Obviously, a sample should not be taken solely from one phase of a multiphase system unless the distribution of the sample between the different phases is known. An example of this type of problem typically arises in oil/water samples. Care must be taken to ensure that specific components of a mixture are not lost during sample 'work-up'. For example, the more volatile components of a mixture can easily be lost if careless weighing procedures are employed. In a similar manner, when solid samples are dissolved, complete dissolution must be achieved or the sample must be weighed beforehand. These points may appear obvious but may not be attended to under the stress of preparing a particularly intractable sample and pressure of time. It must be borne in mind, however, that "what is theoretically ideal may not, in all circumstances, be practically possible". Sometimes compromises must be made to obtain, at best, an approximate quantitative estimation of the components of a particularly difficult sample. Nevertheless, it must be emphasized that quantitative accuracy will not be obtained if the sample is not handled correctly.

If it is found necessary to dissolve the sample in a solvent which is foreign to the material being analyzed, then its stability in that solvent should be ascertained. Particular care should be taken to ensure that the sample does not react with any dissolved oxygen in the solvent and it may be necessary to degas the solvent prior to use with nitrogen or argon. The degassing procedure must be carried out only on the solvent, never the solvent containing the sample, otherwise selective loss of the more volatile components will occur. In some extreme cases it may be necessary to carry out all sample preparation procedures in an inert or anhydrous atmosphere. In the past, many samples not amenable to separation by GC due to their high polarity or low volatility were chemically changed to render them compatible with the technique. This derivatization procedure in GC is less popular now than it used to be due to the alternative choice of the technique of LC for the analysis. Nevertheless, derivatization procedures are still resorted to under special circumstances. Substances such as alcohols, amines and free acids are often derivatized to reduce their interaction with high activity sites on the support or column walls. However, when employing derivatization techniques, care should be taken in all the chemical procedures. It is essential to ensure that there is no loss of sample, that derivatization reactions proceed to completion and any extraction procedures that are used are efficient. An excellent and detailed discussion has been given by Lawrence (1) on derivatization techniques for chromatographic analysis.

As far as the gas chromatograph itself is concerned, care should be taken *not* to inject nonvolatile substances on the column or if necessary employ a guard column to trap such materials. Injection ports can be cleaned and if capillary columns become contaminated, the first foot or so can be broken off and the unclean section removed. This procedure obviously has a limit and if the length of the column is significantly shortened by repeated removal of contaminated sections, the resolution of the column will ultimately become impaired. It is best to avoid contamination in the first place by practicing good chromatographic hygiene. Incidentally thermal decomposition can also easily occur in the heated injection port and consequently the temperature of the injection port should not be greater than necessary as the decomposition products can lead to progressive deterioration of both analytical precision and accuracy.

SAMPLE INTRODUCTION

Syringe Injection

Liquid and gas samples may be introduced into the gas chromatograph by syringe injection or by means of a sample valve, the choice of which depends on the nature of the sample. Syringe injection involves drawing the sample, as a fluid, into a specially designed hypodermic syringe and then inserting the syringe needle through a septum into the injection port of the gas chromatograph. The injection port is often heated above the boiling point of the highest boiling component in order to flash evaporate the sample onto the column. Consequently, samples that are thermally labile often require the injection port to be fitted with a glass liner to reduce the probability of decomposition on the active metal surfaces of the port. Furthermore, residues from prior injections can accumulate and the deposits so formed may eventually have an undesirable effect on the accuracy and precision of the analysis. To reduce the effect of sample accumulation, the injection block should be frequently cleaned and care should be taken to ensure that active substances such as metallic catalysts (samples containing metal particles) are not injected onto the column.

In fact, from both a theoretical point of view and in practice, it is entirely unnecessary to 'boil' the solute mixture onto the column; it is only necessary to bring the sample into equilibrium with the stationary phase at the column temperature. Consequently, if a packed column is employed, the sample need only be injected directly into the column packing with the injection port temperature (if controllable) merely set at the column oven temperature. Even thirty years after the inception of the technique of GC, this is not fully understood. It is still generally thought that chromatographs must have injection ports held at elevated temperature with respect to that of the column and above the boiling point of the highest boiling component of the sample. Only if capillary columns are employed (which are often *not the best choice* for precise and accurate quantitative GC analyses) in conjunction with certain types of injection devices, need high temperature injection ports be resorted to. On-column injection procedures should always be used wherever possible. The on-column injection technique may be especially useful in cases where the sample is sensitive to heat.

The major source of quantitative error when employing syringe injection arises from *inaccuracies* in *delivering* a *precise* and *accurate sample volume* onto the column. Fortunately, errors resulting from this problem may be overcome by the use of an internal standard in the manner discussed in Chapter 3. Another major source of error is the selective evaporation of components from the syringe needle which of course is exacerbated by the use of a high temperature injection port (2). This is especially important when samples that contain components having a wide range of volatilities are being injected. It is a more significant problem when the injection volume is very small relative to the needle capacity. To minimize the problem, the syringe needle should be left in the instrument for a short but reproducible period of time after depressing the plunger. Some syringes have the plunger situated inside the needle and, as the residual needle volume is quite small, the problem may be partially alleviated.

Injection Valves

An injection valve may also be used to introduce the sample into the chromatograph but these are normally only employed for gas analysis. A diagram of a multiport valve is shown in Figure 1. The valve is first turned to the load position and the sample is passed into port A. The sample flows through B into the loop and the excess is lost through ports C and D. When the valve is turned to the inject position the carrier gas entering at E is directed through the loop causing the sample to be swept onto the column through port F. The use of injection valves is rather more specialized than syringe injection and, as already mentioned, they are often used in the analysis of refinery products and, in particular, gas process streams.

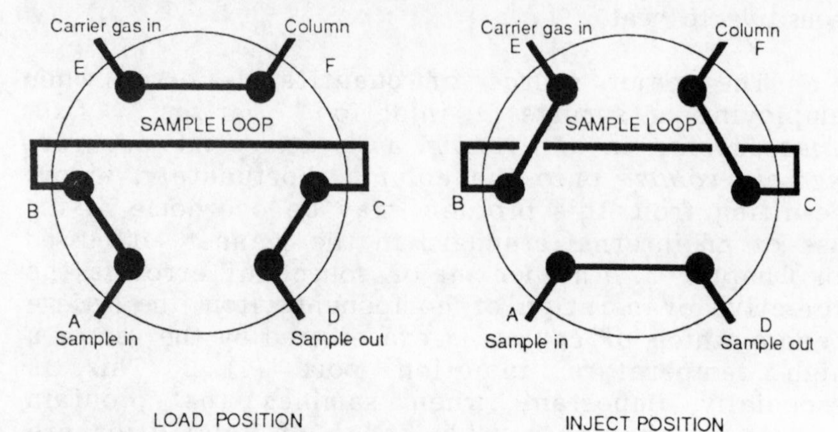

Figure 1. The Multiport Injection Valve.

One area where injection valves have a great advantage is in the *analysis of permanent gases*. The valve does not need to be heated and the loop can be filled at atmospheric pressure. The sample can then be swept out by the carrier gas and, as the sample was taken at atmospheric pressure, the problem of compensating for sample compression does not occur. Syringe injection of small gas samples is not recommended as the dead volume in the syringe needle and connections seriously reduces the accuracy and precision of small injections. Another area where valves are sometimes used is in the *pre-concentration* of samples prior to injection, employing liquid or solid absorption techniques. For example, the valve loop may be packed with an adsorbent such as XAD resin and large volumes of water pumped through the loop (3). The materials of interest are selectively adsorbed on the surface of the resin from the aqueous medium. The water is than displaced from the packed loop by a gas stream and subsequently dried. The loop may then be heated to desorb the sample onto the column. The liquid chromatograph can also be employed as a presample selection or separation technique in conjunction with an appropriate sampling valve (4). This procedure has been given the somewhat vainglorious term LC/GC. The eluent from the LC column is monitored in the usual manner with an appropriate detector and, when the band or bands of interest enters the sample loop, the valve is rotated and the sample is blown onto the column by the carrier gas. In some systems provision is made for removing the LC solvent prior to the separation being developed on the GC column.

Solid Sample Injection

Some specialized analyses require the introduction of solid samples into the chromatograph and there are specially designed injection devices that accommodate such samples (5).

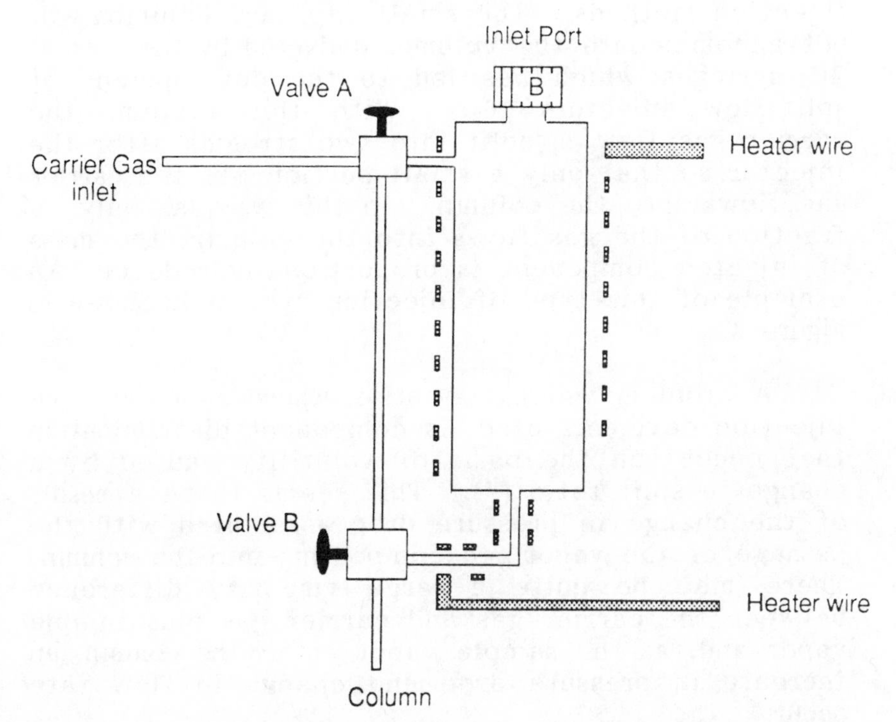

Figure 2. Solid Sample System.

As an example, one may mention the analysis of residual monomer contained in a polymer product. One design for accommodating these samples (shown in Figure 2) has a heated container which can be isolated from the carrier gas. The sample is placed into this container through port B, which is then sealed and the container is heated according to a defined temperature program. After the sample has been heated to the correct temperature and for the required time, the valves are switched so that the evolved vapors are swept directly into the chromatograph.

Split-Flow Injection

Due to their very small loading capacity, capillary columns demand the use of very specialized injection methods. The small capillary columns will not accommodate the volumes delivered by the normal GC syringes which has led to the development of split-flow injectors (6). With this system, the carrier gas flow is split into two streams after the injector so that only a small portion of the carrier gas flows into the column. In this way, as only a fraction of the gas flows into the column, the mass of injected component is proportionally reduced. An example of this type of injection system is shown in Figure 3.

A problem which may arise, whenever a split-type injection device is used, is component discrimination that occurs on the basis of volatility caused by a change in split ratio (7). This seems to be a result of the change in pressure drop associated with the passage of the vaporized components into the column. There may be quite a large viscosity difference between the carrier gas and carrier gas plus sample vapor and, as the sample vapor enters the column, an increase in pressure drop and change in flow rate occur.

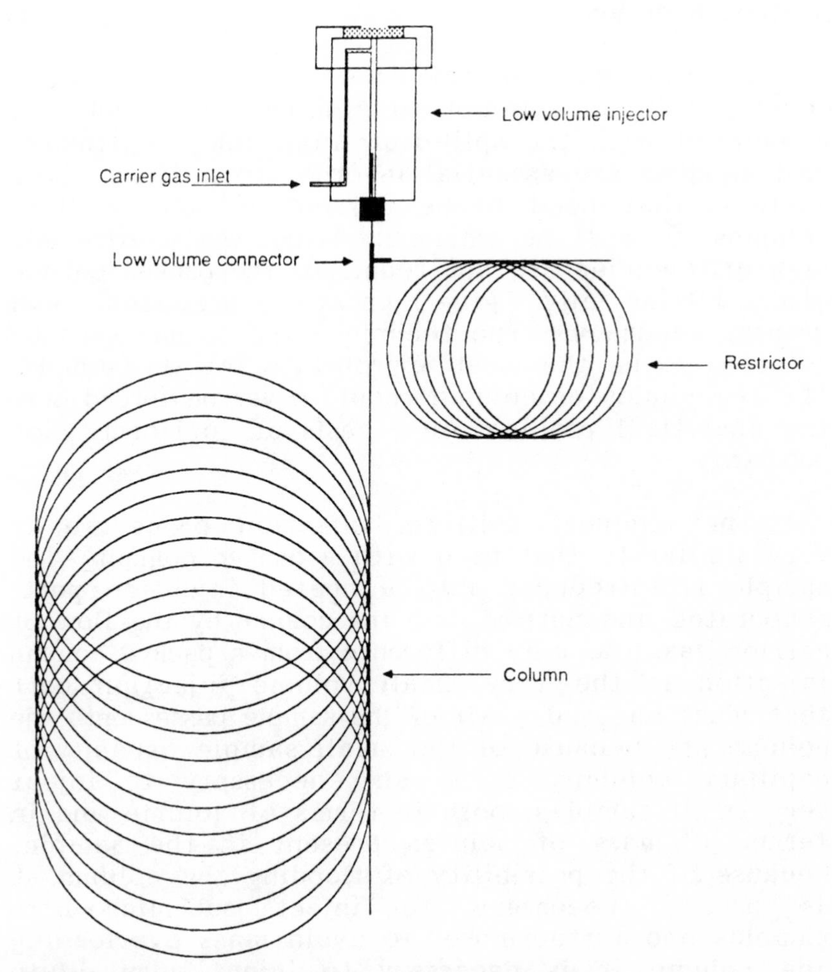

Figure 3. The Split-Flow Injection System for Capillary Columns.

This will cause a change in split ratio as the bypass restricter is not exposed to sample vapor coincident with it entering the column or in a balanced manner. It is thought that the solvent vapor is probably the major cause of this effect.

Splitless Injection

Several splitless injection methods have been developed in an attempt to eliminate the problems associated with the split-flow injection. Although such devices are essential in order to perform those analyses that need to be carried out on capillary columns, it must be emphasized that where extremely high efficiencies are not required, the packed column will, intrinsically give greater precision and greater accuracy. The technique and column must be selected to be the most appropriate for the sample. The techniques or column should never be forced onto the analytical problem on a basis of uniformity or popularity.

The simplest splitless injection procedure is very similar to that used with a packed column. The sample is introduced into a heated injection port, evaporated and carried onto the column by the flow of carrier gas, the only difference from a packed column injection is the very small volume injection port that must be used. All of the sample passes onto the column and because of the small sample capacity of capillary columns, it is also necessary to inject very small samples, both in terms of volume and in terms of mass of solutes present in the sample. Because of the possibility of flooding the column, it is usually necessary to inject sub-micro-litre samples and furthermore, to avoid mass overloading the column it is necessary to inject very dilute samples. Sample concentrations as low as of .01% to .001% are very common. Figure 4 illustrates the principle of the injection method. Figure 4A shows the solute and solvent vapors entering the column, and Figures 4B and 4C show normal distribution behavior where the solvent is proceeding down the column at the carrier gas velocity and the solute is partitioning between the carrier gas and the stationary phase exhibiting a k' value of about 1.

Figure 5 shows how splitting can occur if the sample volume is too large as it enters the column as a liquid. It should be emphasized that Figures 4 to 7 are purely diagramatic and that the injection liquid does not occupy the complete cross section of the tube otherwise the gas flow would be arrested. In Figure 5A, the black rectangle represents liquid sample which has entered the column and in Figure 5B the liquid has split and separated into two zones, a phenomenon which occurs frequently in practice. In Figure 5C, the liquid has evaporated leaving the two zones of concentrated vapor. Figure 5D now shows normal distribution behavior occurring, but there are still two maxima of solute concentration.

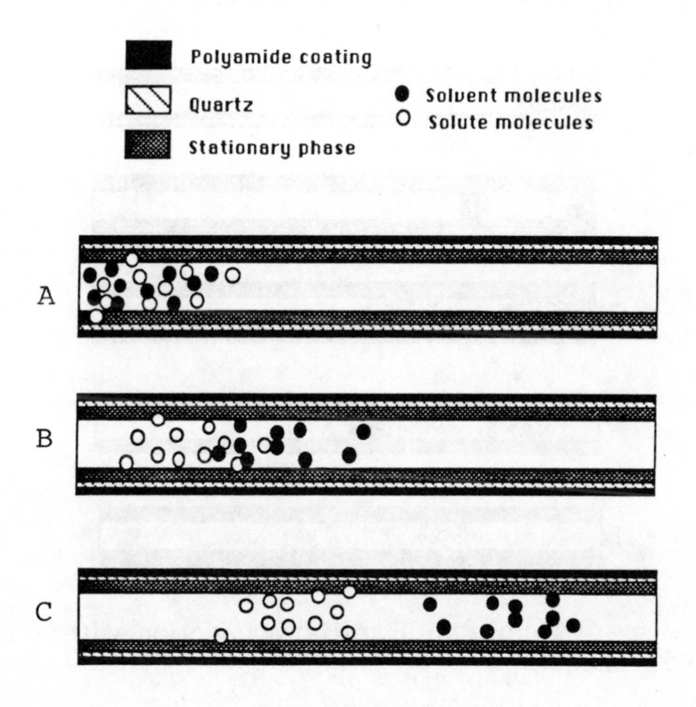

Figure 4. Capillary On-Column Injection.

In Figure 5E it is seen that these zones will stay separated as they move down the column and will exit the column often giving a split peak for each solute component in the mixture on the recorder trace.

Retention Gap Method

A number of techniques have been developed to overcome this problem. In one technique, a 'retention gap' is created at the inlet end of the column (8). This retention gap is a length of column from which the stationary phase has been removed. The retention gap method is depicted in Figure 6A where the white rectangle now represents the region devoid of stationary phase. When the sample enters the column as a liquid, as shown in Figure 6B, it enters this stationary phase-free zone.

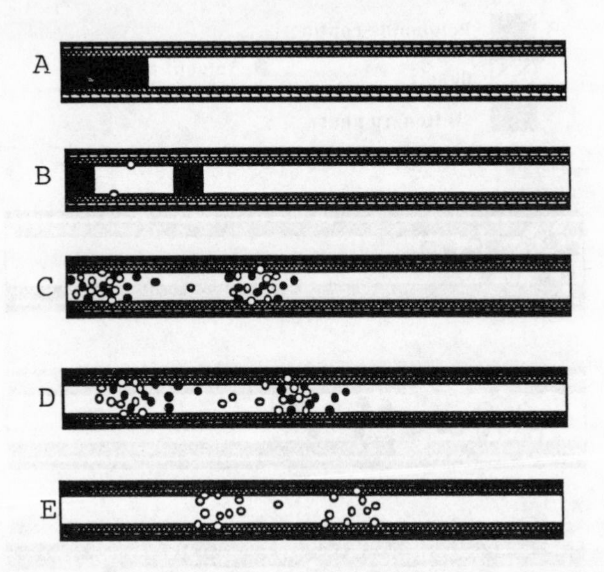

Figure 5. On-column Capillary Injection with a Large Sample Volume.

Now if the sample splits into two zones, as shown in Figure 6C, and then evaporates as shown in Figure 6D, the solute vapors which are in the non-retentive zone will move at the carrier gas velocity, just as the solvent vapors do. However, those solute molecules which have reached the normal column region (which contains stationary phase) will slow down because they partition into the stationary phase. This will allow the second solute zone to catch up to the first, Figure 6E. The efficiency of this process will depend upon the k' of the solute molecules with respect to the stationary phase. For this reason, the column should be temperature programmed and the starting temperature should be fairly low so that the solute molecules are initially well retained on injection. The last diagram in Figure 6F shows normal retention behavior now occurring, but the two concentration zones have nearly merged into one.

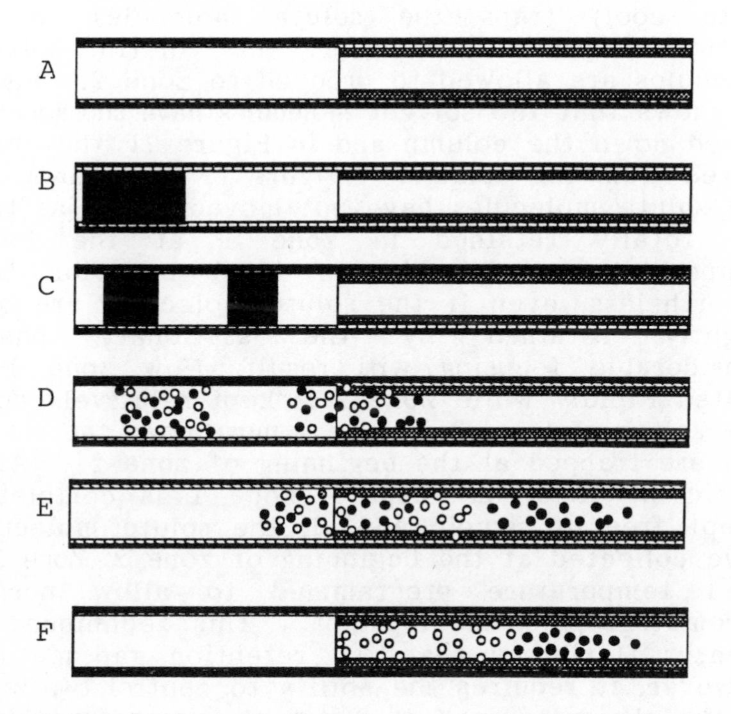

Figure 6. The Retention Gap Method for Sample Injections on Capillary Columns.

How effective this retention gap method is depends upon the difference in boiling point between the solute and solvent molecules, the solvent polarity and the initial program temperature.

Solute Focusing Method

Another technique illustrated in Figure 7 is the solute focusing technique (9). This technique requires a column oven with the ability to heat two zones of the column independently at significantly different temperatures. In Figure 7A, B and C, the injection of the sample and the splitting into two liquid bubbles is seen to take place just as depicted in Figure 5. Figure 7D shows that zone 1 (which is quite cool) traps the solute molecules in its stationary phase. However, the volatile solvent molecules are allowed to proceed to zone 2. Figure 7E shows that the solvent molecules have subsequently moved down the column and in Figure 7F they have exited from the column. Figure 7F also shows that the solute molecules have not moved at all as they are totally retained in zone 1 at the lower temperature. This is an ideal behavior but, nevertheless, even if the solute molecules are only slightly retained by the stationary phase, considerable focusing will result. Now zone 1 is heated rapidly while zone 2 is kept relatively cool. The solute molecules in zone 1 move down the column and are trapped at the beginning of zone 2. After sufficient time has elapsed, zone 1 is completely swept free of sample and all the solute molecules have collected at the beginning of zone 2. Zone 2 is then temperature programmed to allow normal chromatographic development. This technique has greater flexibility than the retention gap method. However, it requires the ability to control two zones of the chromatograph at different temperatures. Most gas chromatographs do not have this capability. Again it should be stressed that this somewhat complicated procedure is entirely unnecessary if a packed column is used which will also provide better precision and accuracy.

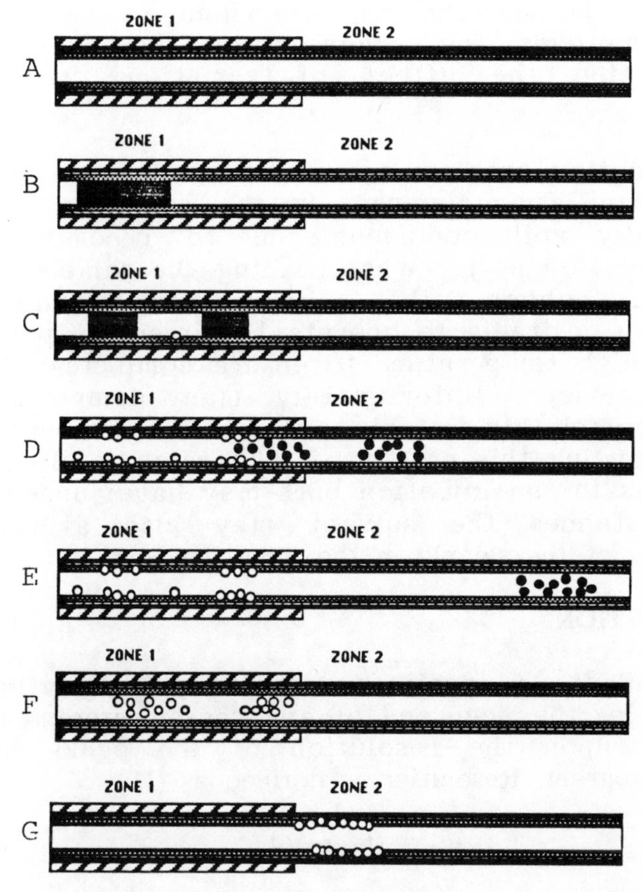

Figure 7. The Injection of Samples onto Capillary Columns Employing Zones of Different Temperature.

Capillary columns are appropriate when the extreme complexity of the sample demands it.

Some of the problems resulting from component discrimination may be alleviated by choosing an internal standard which is similar in volatility to the components that are being analyzed. If a sample containing a number of components having a wide range of volatilities is being analyzed, then it may be desirable to have multiple internal standards. Internal standardization, described in Chapter 3, also improves the accuracy of the analysis by eliminating the effect of variations in sample volume.

Another problem associated with sample injection onto capillary columns is the selective loss of thermally labile components due to decomposition in the injection port. When placing the sample onto a capillary column with a split injection technique, it is often desirable to operate the injection port at a fairly high temperature to ensure complete and *rapid* volatilization. Unfortunately, many compounds are not completely stable at such temperatures. Compounding this problem is the catalytic effect that residues in the injection port may have: under some circumstances the solvent may also affect the stability of the sample in the injection port.

RESOLUTION

Unless a peak deconvolution algorithm is available, the accuracy of any peak measurement will depend upon the resolution of the peaks in the chromatogram. Resolution is defined as (10)

$$R_s = (t_1 - t_2)/4\sigma$$

$$R_s = \Delta t/4\sigma$$

where t_1 is the elution time of the first peak
t_2 is the elution time of the second peak
Δt is the difference in the retention times of the two peaks
and σ is the time standard deviation of the first eluted peak of the critical pair.

The standard deviation for a well resolved Gaussian peak can be determined quite easily by taking advantage of the fact that the width of the peak at .607 of its height is twice the standard deviation (11). These points are illustrated in Figure 8 where two peaks are shown with a separation of 3σ, that is, they have a resolution of .75. The assumption is made that the two peaks have the same time standard deviation as they are eluting close together (12). This will be true provided the two solutes of the critical pair (13) do not differ extremely in molecular weight. The peaks in Figure 8 are of equal height and are shown as if they were in fact separated. Several comments can be made regarding the situation depicted in Figure 8.

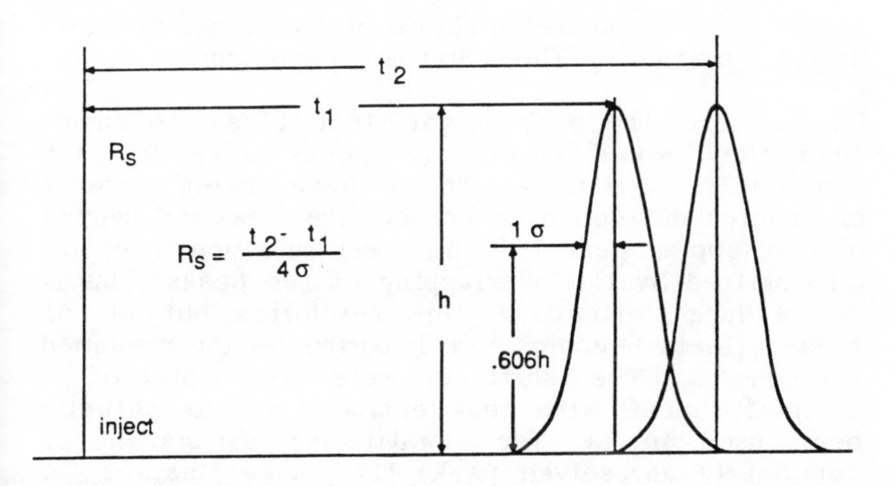

Figure 8. Two Peaks of Equal Height Separated by Three Standard Deviations.

First, the actual peak envelope would look very much like that shown in Figure 9. One can see from Figure 9 that if the overlapping peak is measured at .607 of the peak height, the value of the standard deviation will be approximately 10% too high. This is because the leading edge of the second peak and the trailing edge of the first peak are mutually distorted.

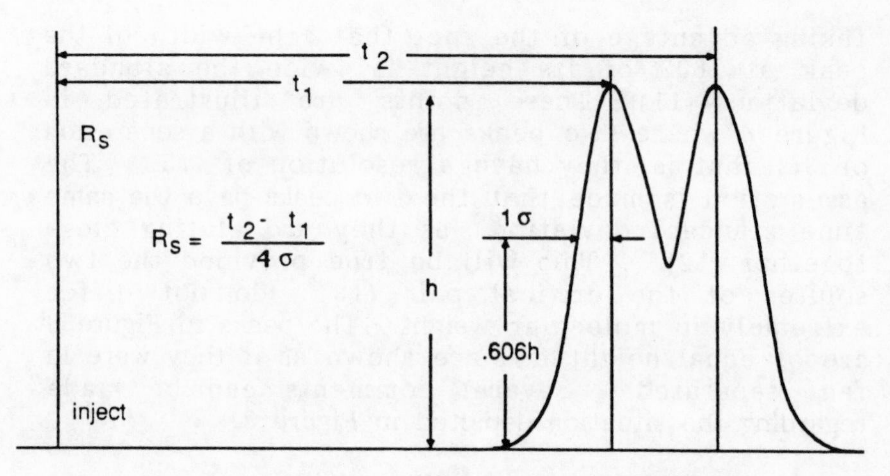

Figure 9. Composite Envelope of Two Peaks of Equal Height Separated by Three Standard Deviations.

Furthermore, the peak heights themselves are higher than they would be if the peaks were in fact completely separated. There have been several procedures developed to correct the areas and heights of overlapping peaks. The peak retention times are also shifted by the overlapping of the peaks. This is not a large effect at this resolution but all of these effects become more important as the resolution decreases. The shift in retention time of a composite peak with peak composition has actually been used in LC for quantitative estimation of completely unresolved peaks (14) (see Chapter 3). The effect of this peak overlap of the form shown in Figure 9 on peak area and height measurement depends upon the method of measurement and will be discussed with the description of the various measurement methods.

There are two methods for increasing resolution in gas chromatography. One is to increase the separation ratio between the two closest eluted components in mixture that are being separated; (the critical pair) (13) and the other is to improve the efficiency of the column so that the peak widths are reduced. The separation ratio of the critical pair can sometimes be improved by lowering the temperature but this is achieved at the expense of increased analysis time. If adequate resolution cannot be obtained by this method without increasing the analysis time unacceptably, then the nature of the stationary phase can be changed to one that exhibits a different selectivity. The efficiency of the column can also be increased by increasing its length. This approach will, unfortunately, also result in increased analysis time (13). In general, the resolution can be expected to increase as the square root of the length of the column and the analysis time directly as its length. Doubling the length of the column will give a 1.4 x increase in resolution and twice the analysis time. An alternate method of increasing resolution by reducing the peak width can be accomplished by operating the column at optimum flow velocity or using particles of smaller size with which to pack the column (15,16). Usually, most commercial columns that are available are packed with the smallest particles commensurate with acceptable flow impedance at the normal inlet pressures available. The use of capillary columns has provided the possibility of very high efficiency (16). The difficulty with capillary columns, however, is the severe limitation placed on sample loading and consequently, the mass range of the analysis.

CHOICE OF DETECTORS

Gas chromatography is fortunate that a sensitive, nearly universal detector is available for all types of analysis. This is, of course, the flame ionization detector (FID) (17). The flame ionization detector is nearly ideal for quantitative analysis as it has a response to solute concentration that is linear over 4 to 5 orders of magnitude (18). It responds to nearly all organic species and it has relatively high sensitivity. The FID response factors do vary with operating conditions and care should be taken to precisely control the hydrogen and air flows (18). It is also necessary to have the detector temperature controlled independently of the column oven. All modern gas chromatographs are well designed with respect to the FID detector requirements and should provide reliable, trouble free operation. The only care the operator must take is to assure that clean gases are supplied and the detector is kept clean. The most common problem is detector contamination from samples which form insoluble non-volatile residues when burned. A good example of this is the silicon dioxide which forms when silanizing reagents are injected to deactivate the stationary phase support material. Another common source of detector residue is from silicone stationary phases when used at too high a temperature. Silicon dioxide may be removed by injecting a Freon type fluorocarbon material that forms HF when burned and then reacts with the SiO_2. This procedure should only be carried out with provision for immediate extraction of all detector vapors from the laboratory. Hydrogen fluoride vapors are very dangerous and should never be allowed to come into contact with laboratory personnel.

The detector which is the second most commonly used is probably the hot wire or katharometer detector (19). This detector senses the difference between the thermal conductivity of a pure carrier gas and carrier gas with added sample vapor. It has the advantage of being a universal detector when used with helium as the carrier gas. However, it is not very sensitive and has a limited response and linear dynamic range. It is commonly used for the analysis of the permanent gases which are not detectable by the FID.

The other common detectors used in GC include the electron capture detector (20), the flame photometric detector (21), the photoionization detector (22) and the nitrogen phosphorus detector (23). These detectors are used in circumstances where their increased sensitivity and/or selectivity confer special advantages. An overview of GC detectors together with their performance characteristics are given in Chapter 4. In general these detectors are suitable for quantitative analysis but may have only a very limited linear range and as a result correction factors may have to be applied.

METHODS OF PEAK AREA MEASUREMENT

In the early days of gas chromatography a major problem was the accurate measurement of appropriate peak parameters. Several manual methods were evolved which gave acceptable results; however, today, electronic integrators and computers have greatly improved the accuracy of peak measurement. As a consequence, manual methods are now infrequently used and only a brief discussion of these non-electronic methods will be given.

A major disadvantage of manual measurements is the necessity that all peaks of interest must be completely contained on the chart recorder. This severely limits the dynamic range of solute composition that may be analyzed. One solution to this problem is to set the attenuator switch to an optimum position before each peak but this can be a tedious and labor-intensive procedure. A second difficulty is making precise measurements on small, narrow peaks. A magnifying reticule or comparator may be used to improve the accuracy of peak width measurements. The best accuracy will be obtained when the standards give responses similar to the unknowns and the chart speed and detector attenuator are adjusted to give large peaks. In GC it is generally thought that peak area measurement techniques are more sensitive to peak overlap than measurement of peak height. Nevertheless, whatever measurement technique is used, the same method must be employed for all the peaks of interest in the chromatogram or suitable corrections must be applied.

Height X Width at Half-Height

The most commonly used manual area measurement technique is obtained by multiplying the peak height by the peak width at half-height. Figure 10 illustrates this method. A line is drawn at the base of the peak and the peak height h is measured. The height is divided by two and a mark is made at that point in the center of the peak. The peak width w is measured at this point. The width should be measured from the inside edge of one side of the peak to outside edge of the other side of the peak. The equation used to calculate the area is as follows:

$$\text{'Area'} = wh$$

This is not in fact the true peak area, but will always represent the same proportion of the true peak area if the peak form is Gaussian.

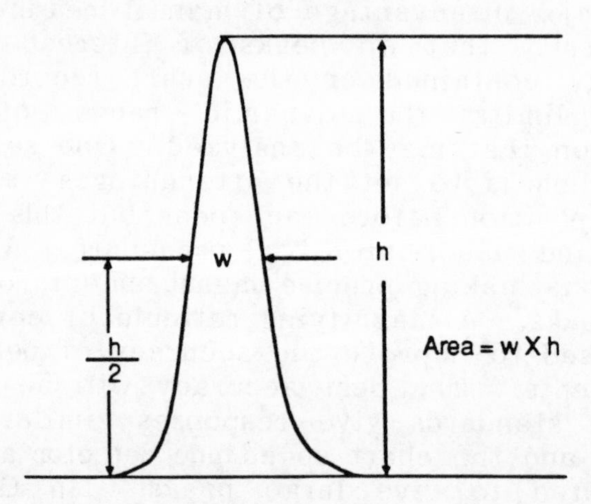

Figure 10. Measurement of Peak Area by the Product of Peak Height and Peak Width at Half Height.

The major source of error with this technique is the difficulty of accurately measuring narrow peaks; this measurement must be as precise as the required precision of the analysis. This problem may be reduced by running the recorder at a speed sufficient to give a peak wide enough to be accurately measured. If significant peak overlap occurs, this method of measurement will give serious errors; the chromatogram shown in Figure 9 will given an error of approximately 10%. If the peaks were not of equal heights, this error estimate would be greater and area values would have to be corrected by a function of the ratio of their heights. Serious errors will also occur if the peaks are significantly skewed. The only advantages of this method are that it is simple and the measurements, themselves, are objectively defined.

The Triangulation Method

Peak triangulation was also one of the early manual methods of peak measurement. In this technique, illustrated in Figure 11 , the tangents to the peak at the inflection points are drawn and the area of the triangle formed by the intersection of these tangents and their intersection with the base line produced is calculated. The area is then given by:

$$\text{'Area'} = w_B/2$$

In practice it is not necessary to divide by two as all of these types of area measurements give an area value which is, at best, proportional to the true area of the chromatographic peak.

A major difficulty with this method is drawing the tangent lines so that they are truly tangential to the inflection points; it is easy for a subconscious bias to affect drawing the tangents.

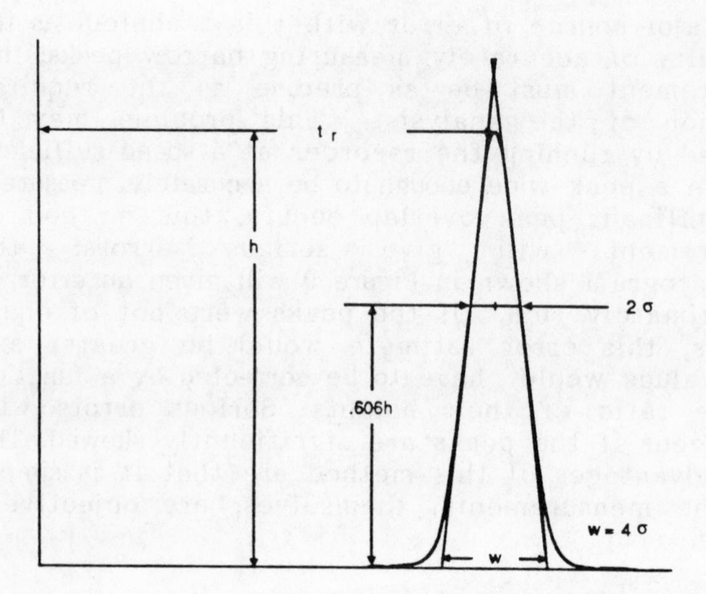

Figure 11. Area Measurement by Peak Triangulation.

Nevertheless, measurement errors tend to compensate to some extent. Any construction errors that reduce the peak height will increase the peak width at the base and *visa versa*. The results involve the same level of error from overlapped peaks as the peak half-height method and even larger errors will result if the peaks are not Gaussian. This method is not recommended; but if used, works best when the peaks are large and similar in shape.

The Planimeter Method

A third method for measuring peak areas is by the use of a planimeter. This is a mechanical device designed to measure the area of any closed plane figure. A pointer is used to trace out the boundaries of a peak. A dial on the device is read before and after tracing out the area and the difference in dial readings gives the area in absolute units. The advantage of a planimeter is that the area of asymmetric peaks can be easily measured. If peaks are overlapped, as shown in Figure 9, a perpendicular can be dropped from the minimum of the valley between the two peaks to the baseline forming a complete perimeter for the two separate peaks and a planimeter will give reasonably accurate results because of the symmetry of the system. Peak of unequal heights will cause errors which can again be reduced by using a correction factor computed from the overlap area and the peak height ratio (24). The major disadvantage is that this method requires a fair amount of manipulative skill and great care is needed to accurately trace the perimeter of the peak. It is also a fairly expensive device when the price is compared with a low cost electronic integrator.

The Cut and Weigh Method

Another method is to cut out and weigh the peaks. The major disadvantages that have been claimed for this method in the past, is that it destroys the chromatogram and most chart papers do not have a very large weight to area ratio and consequently the actual weight of a 'paper peak' is small . If this technique is used, it is recommended that the chromatogram be copied and the peak cut out from the copy. If a heavy bond paper is used then the accuracy of the measurement can be enhanced. It is also possible to enlarge the chromatogram if the peaks are only moderate in size. This is a simple, rapid technique for those cases where only the occasional chromatogram needs to be quantitatively analysed. The major error is in the cutting process, but if the peaks are large and care is taken, the method gives acceptable results with reasonable accuracy and precision. Several copies may be weighed and the results averaged, however, if high accuracy is desired it is best to use one of the electronic integrators. An advantage of the peak weighing method is that it has somewhat more flexibility in dealing with peak overlap than some of the other methods. In this regard it is similar to the planimeter and under such circumstances, unless very sophisticated algorithms are employed, *may* give *more accurate* results than the electronic integrator.

Peak Height x Retention

The final non-electronic area measurement method which will be discussed is the product of the peak height times the retention distance. This method works effectively only with isothermal chromatographic development as it relies upon the assumption that the width of a peak varies linearly with retention time. This relationship is only approximate and this method is best used where many peaks need to be analyzed and high accuracy is not essential. It has the advantage that, as the peak width is not directly measured, it is not as sensitive to peak overlap as the other area measurement techniques. In this regard, it is similar to peak height measurement. If the peaks in Figure 9 were determined with this method only a 1% error would be introduced, if the peaks were not of equal heights this error estimate would have to be corrected by a function of the ratio of the heights. A table of error estimates for different degrees of overlap and peak skewness can be found in reference (25).

PEAK HEIGHT MEASUREMENT

The peak height may be measured and used as an indicator of component mass as most gas chromatographic detectors give a linear response to component concentration. The peak height measurement has the advantage of being simple as it is relatively easy to measure accurately. It is far less sensitive to peak overlap problems than peak area measurement.

The disadvantage is that peak height changes

with retention time and with the operating conditions. If two substances of equal concentration with the same response factors are separated, one expects that the area of the peaks is the same, however, the first eluting peak will be taller than the second. The area depends only on the mass injected but the height depends on the injection volume, the retention time, the column efficiency, the injection technique of the operator, and is non-linear with mass injected if peak overloading is occurring. If, however, operating conditions can be kept sufficiently constant and suitable reference standards are employed, accurate and precise results can be obtained.

ELECTRONIC INTEGRATION

The major development in measurement of GC peaks in the last decade has been the advent of the inexpensive microcomputer. The use of these devices and related inexpensive integrators has all but eliminated the use of manual peak measurement. They extend from systems based on computers costing less than $100.00 to systems costing over $100,000 with something in every price range between. Computer data processing has been discussed in Chapter 2 so only some brief comments will be made here. The major features differentiating these systems are the number of channels, the data acquisition rate, data storage facilities, the reporting speed and capability for integration with other laboratory facilities.

The choice of a system is a complex function of the available budget, the volume of work that will need to be processed, the cost of labor, statutory regulations, etc.

If a multi-channel system is being purchased, the number of channels should be sufficient to enable the simultaneous data acquisition from all the chromatographs in the laboratory. However, it must be remembered that as the number of channels increases and the time saving capacity also increases, more terminals are required to handle the workload. However, it must also be said, that many workers prefer to have a number of smaller data acquisition systems with two to four channels per system so that system breakdown only effects a small proportion of the equipment. The sampling rate should be sufficient for the type of chromatography to be used: for normal packed column work, a data rate of 10 samples per second is sufficient.

PROCESSING THE DATA FOR QUANTITATIVE ANALYSIS

There are three general methods of computing the analytical results and they have already been discussed in Chapter 3. They are Normalization, External Standardization and Internal Standardization. Further brief comments will be given here that relate these methods specifically to GC analysis. The area normalization method assumes that all components are eluted and that all components have the same response factor. In practice, and in contrast to LC analysis, these conditions are never met, however, the response factors for some components can be quite similar on some detectors. One may mention, for example, the response factors of the higher molecular weight saturated hydrocarbons (found in paraffin waxes) for the FID detector. The accuracy of the analysis will deteriorate if the assumptions are invalid but this method gives satisfactory results in this case and it is very easy to use as no standards are necessary . To compute the analysis, the areas obtained for the peaks are summed and the concentration of each peak is given as the fraction of its area to the total area of all the peaks.

The external standardization method is most efficient if the pure standards of the peaks of interest be available. Separate standard solutions of known concentration of these solutes are made up and chromatographed. Several injections of different masses of the standard should be chromatographed so that response factors over the range of concentrations expected in the unknown sample may be computed. It may be necessary to make up several calibration solutions if a wide range of solute concentration is to be covered so as to avoid volume overloading the column. This would result in different vapor mass concentrations being seen by the detector which may, over extreme concentration ranges, be non-linear. A response factor (ψ) for each standard may be calculated from the following equation:

$$\psi = (\text{volume injected})(\text{standard conc.})/(\text{peak area})$$

The response factor will be given in units of mass/unit area or mass/unit height. The sample is now chromatographed under the same conditions and the mass of each component injected is computed by multiplying its area by the response factor of its standard. The concentration in the sample solution is obtained by dividing the mass by the injection volume. This method compensates for varying response of the detector for different solutes and it does not require that all the solutes are eluted. It does require that a known volume of sample be injected and this is the major source of short term error with this method. It is also necessary to re-run the calibration standard solutions frequently between a series of analyses, particularly if the operating conditions are likely to have changed. Large changes in response factor may result from changing the hydrogen flow to the FID detector for example. If peak heights are being measured, then care must be taken to ensure that the chromatographic conditions such as column temperature and flow rate are also kept constant.

The third method is the internal standard method. In this case the reference or standard must be completely separated from all other components in the mixture. A mixture is made up consisting of a known concentration of the standard and those substances of interest in the mixture. For simplicity only one component of interest will be considered. The mixture is chromatographed and the area of the standard and the solute of interest measured. Reiterating the equation given in Chapter 3, but in a slightly different form, if c_{st} and A_{st} are the concentration and peak area of the standard, respectively:

$$F_{st} = c_{st}/A_{st}$$

In a similar way if c_{int} and A_{int} are the concentration and area of the solute of interest, then:

$$F_{int} = c_{int}/A_{int}$$

Combining the two above equations:

$$f = F_{st}/F_{int} = c_{st}A_{int}/c_{int}A_{st}$$

Now if a mass of standard is added to an aliquot of the sample to provide a concentration of c'_{st} and the sample chromatographed to provide peak areas of the solute of interest and standard of A'_{int} and A'_{st}, respectively. Then the required concentration of the solute of interest c'_{int} is given by :

$$c'_{int} = fA'_{int}c'_{st}/A'_{st}$$

It is to be noted that this equation does not contain any reference to the volume. This is the major advantage of the internal standard method; it is also less sensitive to changes in detector response factors as most changes will affect the sample component and internal standard in the same way resulting in only small changes in the value of *f*.

SYNOPSIS

Precise and *accurate analysis* in GC requires *adequate resolution*. The *sample must be representative* of the material being analysed and care must be taken to ensure that the sample does not incur *loss by oxidation* or selective *evaporation*. Good chromatographic hygiene should be practiced, dirty samples should not be injected onto the column and the injection port and associated apparatus should be regularly cleaned. *Excessive inlet temperatures should be avoided* to restrict thermal decomposition and this can be aided by the use of on-column injection. A major source of error results from variable injection volumes due to poor injection techniques. Selective evaporation of the more volatile components is also exacerbated by the use of elevated injection port temperatures. *Injection valves* are best used for the *analysis of gas samples*, but if a packed loop is employed, then a sample concentration process is possible. Syringe injection is not recommended for gas samples. Valving procedures can be used for the analysis of volatile materials in solid samples, e.g. the determination of monomers in polymers. The *low sample capacity of capillary columns demand special injection techniques*. As a result, a number of injection procedures for capillary columns have been devised: *split injection*, where only a fraction of the sample passes onto the column (this procedure tends to discriminate on the basis of volatility); *splitless injection* of very small samples at very low concentration in the solvent (this procedure tends to give a very limited concentration range for analysis and can cause 'double peaking'); *the retention gap method* where the sample is placed onto a section of the column which is not coated (this procedure improves both accuracy and precision); and finally the *solute focussing method* that requires two sections of the beginning of the column to be capable of being thermostated at different temperatures (this procedure probably gives the most precise and accurate results as far as capillary columns are concerned). When a heated injection port is employed, as with split injection systems, particular care should be taken to avoid thermal and catalytic decomposition of the sample by operating the port temperature at the minimum possible temperature.

Optimum resolution is obtained when *each peak exhibits baseline separation* and this is achieved by either choosing a stationary phase to provide adequate selectivity or a column with adequate efficiency. The former conserves the analysis time, the latter tends to extend the analysis time as it often requires the use of a longer column. The maximum column efficiency is achieved by operating the column at its optimum mobile phase velocity which also extends the analysis time. The resolving power of a column is measured by the magnitude of the ratio of the distance between the peaks to the peak width of the first of the pair. *Poor resolution results* in a *reduced accuracy and precision* both in the measurement of peak areas, peak heights and retention times. The most *common detector* used in GC for quantitative analysis is the *FID* followed by the *TCD* for the analysis of *permanent gases*. The FID combines *high sensitivity* with *wide linear dynamic range*. The detector connecting tubes, electrodes and detector body should be frequently cleaned and kept free from pyrolysis contaminants. Special detectors with higher sensitivity but perhaps a more restricted linear dynamic range should be selected when the nature of the sample demands it. There are *six manual methods* of peak measurement, *Height times Width at Half Height, Peak Triangulation, Measurement by Planimeter, Cut and Weigh Method, Peak Height times Retention* and *Peak Height Measurement*. All manual measurements are in most cases *inferior in accuracy and precision* to electronic integration methods whether by computer or by integrators. There is a wide selection of electronic devices for chromatographic data processing. When selecting a data processing system, it should provide processing facilities for the required number of chromatographs, adequate data acquisition rates, appropriate algorithms for peak measurements and satisfactory reporting formats and methods. There are three basic methods for processing data for quantitative analysis, the *Normalization Method*, The *External Standard Method* and the *Internal Standard Method*. The Normalization Method is rarely employed because it requires that the detector gives the same response to all solutes. The Internal Standard Method gives the most precise and accurate results but requires the selection of a standard that is resolved from all the sample constituents. Furthermore, the solutes of interest cannot, themselves, be selected as standards. The External Standard Method is a little less precise and accurate than the Internal Standard Method but allows the solutes of interest, themselves to be used as standards and makes no extra demands on the chromatographic resolution.

References

1. J.F. Lawrence, in "Chemical Derivatization in Analytical Chemistry", R.W. Frei and J.F. Lawrence (Eds.), Plenum Press, New York (1982) p. 191.

2. K. Grob, Jr., and H.P. Neukom, J. High Resolut. Chromatogr. Chromatogr. Commun., 2 (1979) 15.

3. J.P. Ryan and J.S. Fritz, J. Chromatogr. Sci., 16 (1978) 488.

4. T.V. Raglione, N. Sagliano, Jr., T.R. Floyd and R.A. Hartwick, LC-GC Magazine, 4 (4) (1986) 328.

5. R. Steichen, Anal. Chem. 48 (1976) 1398.

6. D.H. Desty, A. Goldup and B.H.F. Wlyman, J. Inst. Petrol., 45 (1959) 287.

7. G. Schomburg, H. Behlau, R. Dielmann, F. Weeke and H. Husman, J. Chromatogr., 142 (1977) 87.

8. K. Grob, Jr., J. Chromatogr., 237 (1982) 15.

9. J.V. Hinshaw, Jr. and F.J. Yang, J.High Resolut. Chromatogr. Chromatogr. Commun., 6 (1983) 554.

10. J.H. Purnell, No. 4804 Dec. (1959) 2009.

11. R.P.W. Scott, "Contemporary Liquid Chromatography", Wiley, New York, 1976, p. 43.

12. R.P.W. Scott, "Contemporary Liquid Chromatography", Wiley, New York, 1976 p. 36.

13. "The Science of Chromatography", F. Bruner (Ed.) Elsevier, Amsterdam, 1985 p. 403.

14. R.P.W. Scott and C.E. Reese, J. Chromatogr., 138 (1977) 283.

15. "Gas Chromatography 1960", R.P.W. Scott (Ed.), Butterworths, London, 1960, p. 144.

16. K. Ogan and R.P.W. Scott, J. High Resolution Chromatography, 7 (1984) 382.

17. I.G. McWilliam and R.A. Dewar in "Gas Chromatography 1958", D.H. Desty (Ed.), Butterworths, London, 1958, p. 142.

18. I.A. Fowlis, R.J. Maggs and R.P.W. Scott, J. Chromatogr., 15 (1964) 471.

19. J. Sevcik, "Detectors in Gas Chromatography", Elsevier, Amsterdam, 1975, p. 39.

20. J. Sevcik, "Detectors in Gas Chromatography", Elsevier, Amsterdam, 1975, p. 72.

21. J. Sevcik, "Detectors in Gas Chromatography", Elsevier, Amsterdam, 1975, p. 143.

22. J. Sevcik, "Detectors in Gas Chromatography", Elsevier, 1975, p. 123.

23. J. Sevcik, "Detectors in Gas Chromatography", Elsevier, Amsterdam, 1975, p. 105.

24. J.C. Bartlet, D.M. Smith, Can. J. Chem., 38 (1960) 2057.

25. A.W. Westerberg, Anal. Chem., 41 (1969) 1770.

Quantitative Analysis using
Chromatographic Techniques
Edited by Elena Katz

Chapter 6

QUANTITATIVE THIN-LAYER CHROMATOGRAPHY

Colin F. Poole and Salwa Khatib

INTRODUCTION

Thin-layer chromatography (TLC) is presently undergoing a resurgence of interest hurried along by new developments in stationary phases and equipment for sample application, plate development, and in situ quantitation. These new methods are variously called high performance thin-layer chromatography (HPTLC) or instrumental thin-layer chromatography to distinguish them from the less rigorous and qualitative methods that they have replaced (1-6). These newer methods provide higher resolution with shorter analysis times, greatly improved sample detectability, and much greater reproducibility of all aspects comprising the TLC process.

One of the early milestones in the development of modern TLC was the commercial availability of new plates prepared with a homogeneous layer of small particles (5 to 15 μm) of a narrow particle size distribution. The first commercially available HPTLC plates were coated with silica gel (7); subsequently, other phases became available including alumina, cellulose (8), aminopropylsilanized silica (9), reversed-phase plates containing ethyl-, octyl-, octadecyl-, and diphenylsilanized silica (10), and most recently a chiral phase for the separation of enantiomers (11). These new plates have low sample capacities; typical sample volumes applied to the plates are on the order of 100-200 nL to give starting spots of 1.0-1.5 mm in diameter. To insure accurate spot location for scanning densitometry, mechanical sample application devices became essential (2,6). At a later date sample streaking and solid phase transfer devices for overcoming the sample volume limitations imposed by the use of dosimeters were developed (12). These were important in providing a bridge between sample sizes encountered in routine analytical laboratories, the sensitivity needed in many current assays, and the properties of the new HPTLC plates. Automated sample application devices are now commercially available and are just as easy to use as autosamplers for gas and liquid chromatography.

Initially, because of the slow rate of solvent migration through the new plates, solvent migration distances were kept comparatively short, 3-6 cm. However, these short migration distances were more than adequate to provide greater effective sample resolution than could be obtained on conventional coarse-particle TLC plates. The time needed for a particular analysis was also substantially shorter on the new plates in spite of the lower mobile phase velocities because only short migration distances were required. If the mobile phase velocity is controlled by external means, as in forced-flow TLC, then the restrictions imposed by solvent migration using capillary action are removed and both the plate length and mobile phase velocity for a particular separation can be optimized independently of other experimental considerations (13,14). This required the construction of special development chambers in which the sorbent layer was sandwiched between a glass backing plate and an impermeable polymeric membrane forced into intimate contact with the layer by application of hydraulic pressure (15).

The rapid solvent migration for short distances renewed interest in multiple development techniques for the separation of complex mixtures (2). This technique takes advantage of the spot reconcentration mechanism to minimize band broadening in the direction of migration and is easily combined with step solvent gradients for separating mixtures spanning a wide polarity range. Continuous development techniques have been optimized by consideration of the primary experimental parameters, plate length, time, and solvent strength. Continuous development provides a convenient method of improving the resolution of difficult-to-separate components by optimizing solvent selectivity (16).

During the same period that the above changes were taking place, quantitation of thin-layer chromatograms became more convenient and reliable due to the introduction of automated scanning densitometers. The instrumental parameters that affect the observed resolution and signal-to-noise ratio for slit-scanning densitometers are now well established (5,17). Spurious noise due to inhomogeneous scatter from layer imperfections was much reduced with the introduction of the more densely coated and homogeneous HPTLC plates. Spots tend to be both smaller and regular in shape in HPTLC which increases their detectability for similar mass and also the reliability of their quantitation in linear scans. For the same reason the limited sample capacity of the HPTLC plates is not a problem as in practice detection limits are frequently an order of magnitude lower on HPTLC plates compared to coarse-particle plates. Detection limits, for favorable cases, are in the low nanogram range and low picogram range for absorption and fluorescence, respectively, on HPTLC plates. An additional advantage of the compact starting spots typical of HPTLC is that it allows a greater number of samples to be applied along the edge of the plate compensating for the higher cost of the new plates.

Today it is often possible to perform separations of complex mixtures by HPTLC that only a few years ago would have been considered the preserve of high performance liquid chromatography (HPLC). It is not our intention, however, to extol the relative virtues of HPLC and HPTLC in this chapter. In the authors' opinion, HPTLC and HPLC are competitive and complementary techniques, and it is often possible to demonstrate the superiority of one technique over the other, depending on the situation. It is most likely that HPTLC will be favored for the separation of mixtures that are not too complex (containing from 5 to 15 components or thereabouts), for the separation of relatively crude samples of unknown origin, for the separation of samples that contain slow or non-eluting matrix components, or in instances where there is a need for a high sample throughput of samples of a similar kind.

SAMPLE SEPARATION

The stationary phase in TLC consists of a thin layer of sorbent coated on an inert, rigid backing-material. The sample is applied to the surface of the sorbent layer as a spot or band, a short distance from the bottom edge of the plate in *linear* chromatography, on the circumference of a circle at the center of the plate in *circular* chromatography, or in a circle at the periphery of the plate in *anticircular* chromatography. The separation is carried out in an enclosed chamber by contacting the layer behind the sample with the mobile phase, which advances through the stationary phase by capillary forces, or it is forced through the stationary phase by a mechanical pump. A separation of the sample results from the different rates of migration of the sample components in the direction travelled by the mobile phase. After development and evaporation of the mobile phase, the separated components are determined by *in situ* scanning densitometry.

Chromatographic Performance of a Thin-Layer Plate

The fundamental parameter used to characterize

the position of a spot in a TLC chromatogram is the retardation factor, or R_F value. It represents the ratio of the distance migrated by the sample compared to that traveled by the solvent front

$$R_F = \frac{Z_{substance}}{Z_{mobile\ phase}} = \frac{Z_s}{Z_f} \tag{1}$$

where Z_s is the distance migrated by the sample (spot) from its origin
and Z_f is the distance moved by the mobile phase from the sample origin to the solvent front.

Boundary conditions: $1 > R_F > 0$

The capacity factor (k') and the R_F value are related by equation (2)

$$k' = \frac{1 - R_F}{R_F} \tag{2}$$

Separations performed with fine-particle layers are characterized by a series of symmetrical, compact spots uniformly increasing in diameter with increasing migration distance (18). For capillary flow controlled conditions and fine-particle layers, zone broadening is controlled by particle size and by molecular diffusion, while mass transfer contributions are negligible. For coarse-particle layers, elongated and irregularly shaped spots are not uncommon and here the contribution of mass transfer kinetics to spot broadening cannot be ignored. The distribution of sample within a spot is essentially Gaussian for a fine-particle layer and the efficiency of the chromatographic system, determined as the number of theoretical plates, is adequately described by equation (3)

$$n = 16 \left[\frac{Z_s}{W_b} \right]^2 = 16 \left[\frac{R_F Z_f}{W_b} \right]^2 \qquad (3)$$

where n is the number of theoretical plates,
Z_s is the migration distance of substance *s*
and W_b is the spot diameter of substance *s*.

It should be noted that equation (3) contains a term *dependent on the migration distance*, and is thus not a constant term for the layer. By convention, the efficiency of a thin-layer bed is measured or calculated for a substance having an R_F value of 0.5 or 1.0, or an average value is used (18).

An empirical relationship between the average plate height ($\bar{H}$) and the solvent migration distance for coarse- and fine-particle layers with capillary and forced-flow solvent migration is shown in Figure 1 (15, 19). When the migration distance is short, the efficiency of the HPTLC plate (A) exceeds that of the coarse-particle plate (B), for capillary flow controlled conditions (20). For capillary flow controlled conditions, as the solvent migration distance increases, the performance of the fine-particle layer (A) declines until the curve representing the average plate height crosses that of the coarse-particle layer (B). At longer migration distances the efficiency of the fine-particle layer is diffusion limited due to the low migration velocity of the mobile phase. Thus, for optimum separations using HPTLC plates, the development distance should be kept short and, if required, multiple developments over short distances used to improve resolution. Under forced-flow conditions the average plate height is approximately constant and the plate number increases linearly with migration distance. At all migration distances, Figure 1, the fine-particle layer (D) shows the highest efficiency with an average plate height value of approximately 12 μm.

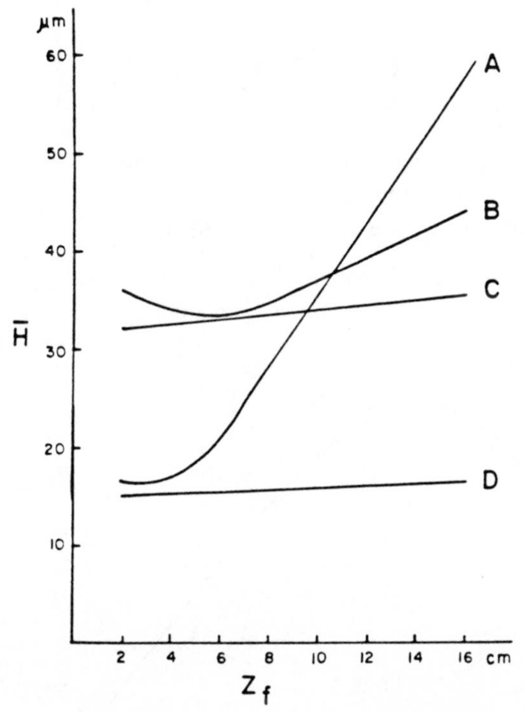

Figure 1. Variation of the efficiency of fine- and coarse-particle layers as a function of migration distance and development technique. A, HPTLC plate, normal development; B, conventional TLC plate, normal development; C, conventional TLC plate, forced-flow development; and D, HPTLC plate, forced-flow development.

The ultimate efficiency of the forced-flow system is limited only by the particle size and homogeneity of the bed and by the pressure required to maintain the optimum mobile phase velocity. Using commercially available equipment and plates, an upper limit of about 50,000 theoretical plates seems reasonable.

The TLC analog of the classical resolution

equation for column chromatography is given below (1)

$$R_s = \frac{1}{4}\left[\frac{k'_1}{k'_2} - 1\right]\left[\frac{n}{\bar{k} + 1}\right]^{1/2}\left[\frac{\bar{k}}{\bar{k} + 1}\right] \quad (4)$$

where k'_1 and k'_2 are the capacity factors constant for spots 1 and 2
and $\bar{k}$ is the mean value of k'_1 and k'_2.

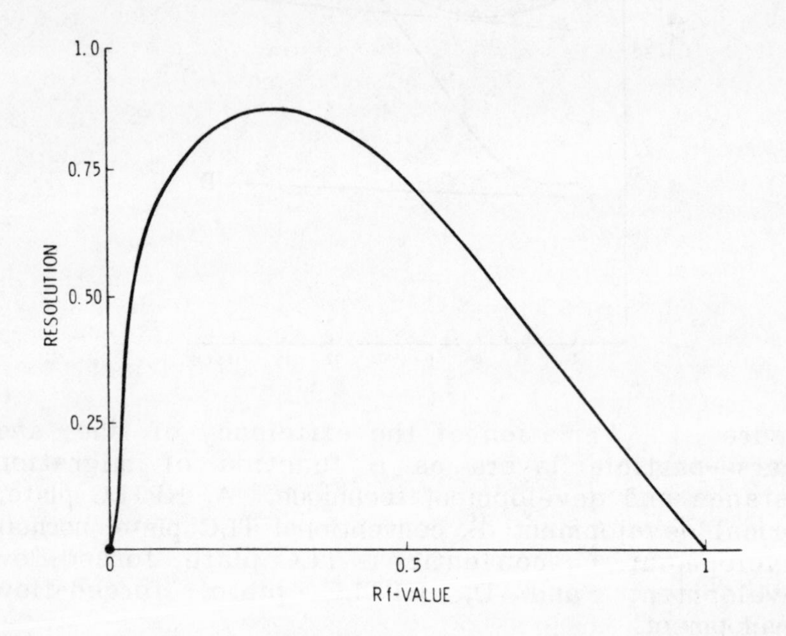

Figure 2. Change in resolution of two closely migrating spots as a function of the R_F value of the faster moving spot.

The first term in equation (4) measures the selectivity of the system, the second the quality of the layer, and the third the relative position of the spots in the chromatogram. Figure 2 illustrates the relationship between resolution for two closely migrating spots as a function of the R_F value of the faster moving component (6). It is seen that the resolution curve is in fact bell shaped with a maximum at an R_F value of about 0.3. The resolution does not change significantly for R_F values in the range 0.2 to 0.5; within this range the resolution is greater than 92% of the maximum value (75% between R_F = 0.1 to 0.6). Outside of this range the resolution declines rapidly illustrating the strong correlation between separating power in TLC and the position of zones in the chromatogram. Under forced-flow development conditions there is no maximum for resolution as a function of migration distance (14). Here the resolution continues to increase as the migration distance increases.

Continuous Sample Development

The separating power of thin-layer chromatography can be improved by using either *continuous* or *multiple* development techniques (2). Continuous development is most useful for improving the resolution of difficult-to-separate component pairs by optimizing the selectivity of the chromatographic system. In the continuous development mode, the mobile phase traverses the TLC plate to a fixed predetermined position on the plate at which point it is continuously evaporated. Evaporation usually occurs at the plate atmospheric boundary using either natural or forced evaporation. During the first part of the development capillary forces are responsible for movement of the solvent front, once it reaches the boundary, additional forces are applied by the evaporation of the solvent. Eventually a steady state constant velocity is reached at which the mass of solvent evaporating at the boundary is equivalent to the mass of new solvent entering the layer. By minimizing the plate length used for the separation, the mobile phase velocity remains high and under these conditions the selectivity of the separation system can often be improved by reducing the polarity of the developing solvent without resorting to long analysis times. Nurok (21) has shown that the minimum analysis time for a given spot separation will always be shorter by continuous development than by conventional development, when the experimental parameters are correctly optimized.

Multiple Sample Development

In multiple development, the TLC plate is developed for some selected distance, the plate removed from the developing chamber and adsorbed solvent evaporated before returning the plate to the developing chamber and repeating the development process. This is a very versatile strategy for separating complex mixtures, as the primary experimental variables (plate length, time of development, if continuous development is used, and composition of the mobile phase) can be changed at each development step and the number of steps varied to obtain the desired separation. Quantitative measurements can be made at several steps in the sequence and, therefore, it is not necessary for all components to be separated at one time provided that they can all be resolved at different segments in the development sequence. A device for automated multiple development recently became available (22).

The flexibility of the multiple development process contributes considerably to its success but, from a theoretical point of view, this leads to the introduction of many variables that are difficult to control. With a better understanding of the TLC process itself, and methods and apparatus designed to provide more rigorous control over experimental variables, such as are used for forced-flow TLC, a predictive model should be feasible. However, as multiple development constitutes one of the most powerful separation tools in modern TLC, a phenomonological sketch of the method will be given in the following paragraphs.

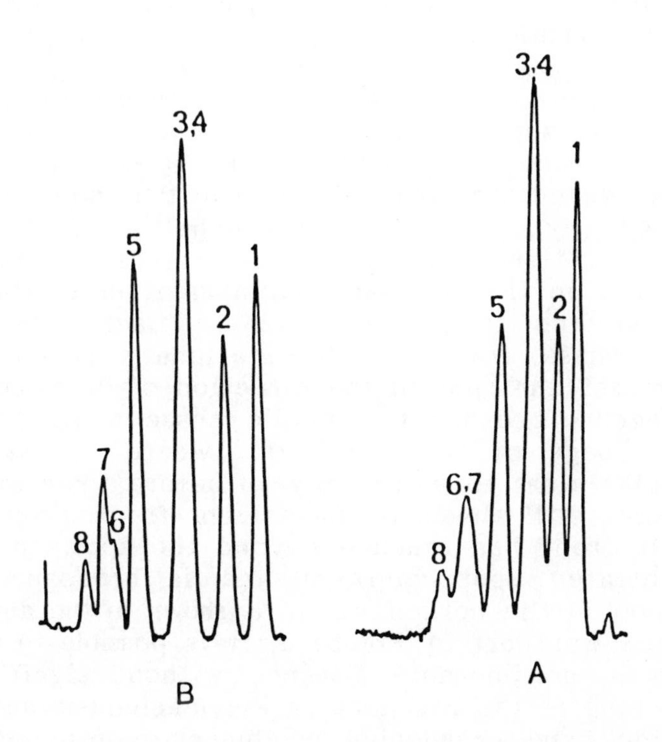

Figure 3. Comparison of conventional (A) and multiple development (B) for the separation of a mixture of polycyclic aromatic hydrocarbons. For multiple development, 4 x 5 min. and 3 x 7 min. developments were used with the mobile phase methanol-water (4:1). The total development time for (A) was the same as for (B), 41 min. Peaks:1 = Coronene; 2 = Benz[g,h,i]perylene; 3 = Perylene; 4 = Benzo[a]pyrene; 5 = Benz[a]anthracene; 6 = Pyrene; 7 = Fluoranthene; 8 = Anthracene.

Figure 3 illustrates the separation of a mixture of polycyclic aromatic hydrocarbons by normal and multiple development TLC. In both cases the time of development, mobile phase composition, and stationary phase are identical.

The separation obtained by multiple development is clearly superior with earlier peaks exhibiting baseline resolution and later peaks partially separated and sharpened compared to the chromatogram obtained by normal development. The spot reconcentration mechanism that leads to higher efficiency in the multiple development chromatogram is illustrated in Figure 4 (23). Each time the mobile phase traverses the stationary sample, it compresses the spot in the direction of development. This occurs because the mobile phase contacts the bottom edge of the spot first, where the sample molecules start to move forward before those sample molecules still ahead of the solvent front. Once the solvent front has reached beyond the spot, the re-concentrated spot migrates and is broadened by diffusion in the normal way. As shown in the diagram in the lower part of Figure 4, it is possible to move a spot a considerable distance without significant broadening if the reconcentration mechanism and the opposing band broadening mechanism approximately cancel each other. The net result is that the spot centers are separated in proportion to their migration distance, whereas the spot widths are hardly changed and, thus, the resolution is much improved.

Multiple development is an important method for the separation of mixtures containing solutes having a *wide polarity* range for which no single solvent system exists for their separation. Each segment of the development process can be independently optimized to provide the desired resolution of just a few components at a time which are then determined by scanning densitometry at the development segment most

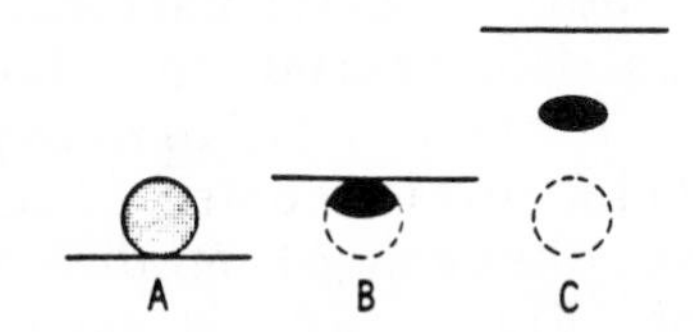

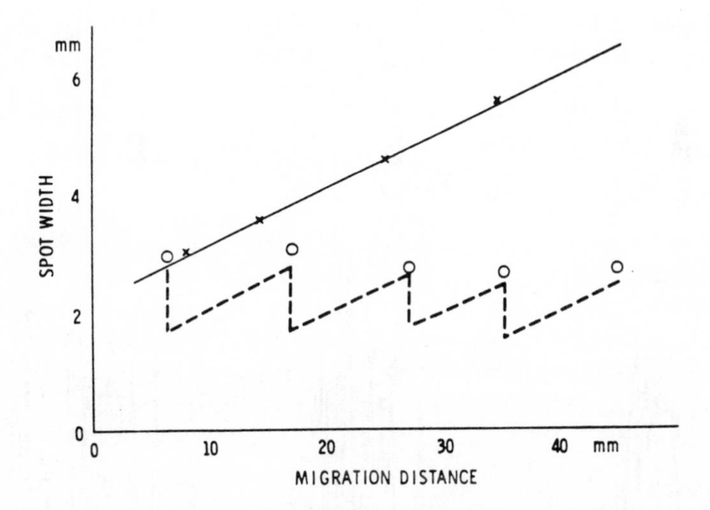

Figure 4. Top: schematic representation of the spot reconcentration mechanism in multiple development. A, advancing solvent front contacts lower edge of spot; B, solvent front traverses spot, producing a compression in the bottom-to-top direction; C, spot is developed normally after being reconcentrated. Bottom: Spot broadening as a function of migration distance. The solid line represents the constant rate of spot broadening by normal development. The broken line illustrates the effect of the spot reconcentration mechanism as a means of controlling spot broadening.

appropriate for their determination. The power of this procedure is illustrated in Figure 5 for the separation of the protein PTH-amino acid derivatives on silica gel HPTLC plates using three development segments and two changes of mobile phase (24). The first development, Figure 5A, is made with methylene chloride for five minutes using a plate length of 3.5 cm. This step is performed to order the PTH-amino acid derivatives at the origin to enhance their resolution in later segments.

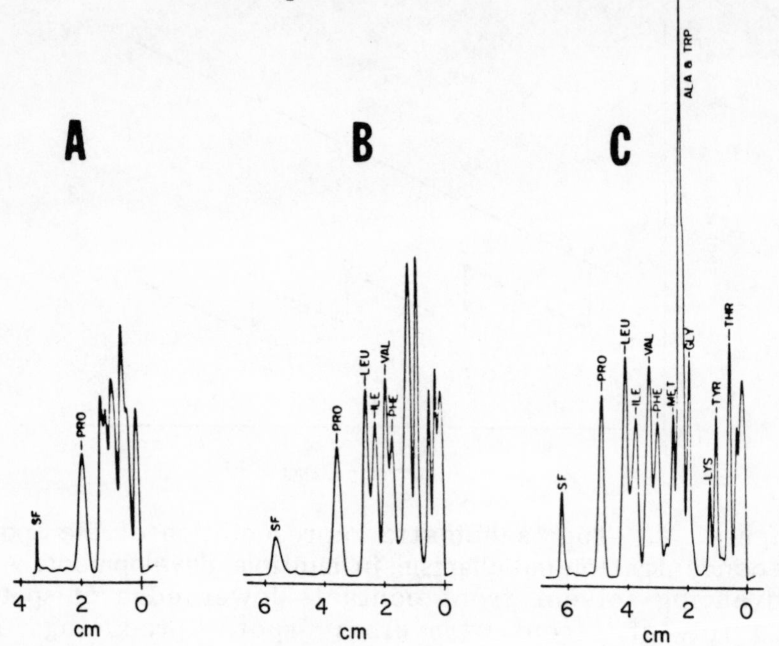

Figure 5. Separation of protein amino acid PTH-derivatives using multiple development with changes in the mobile phase composition (see text for details). The stationary phase was silica gel 60 HPTLC plate.

Although PTH-proline exhibits baseline resolution from other PTH-amino acids in this segment, there is no need to make any determinations at this point as it remains baseline resolved in the next development steps. After the methylene chloride has been evaporated from the plate, it is redeveloped for 10 min in methylene chloride-isopropanol (99:1) with a plate length of 7.5 cm. In the subsequent segments the plate length was unchanged. At this development step, Figure 5B, the PTH derivatives of proline, leucine, isoleucine, valine, and phenylalanine are separated. The third development segment is a repeat of the second, Figure 5C, and provides a better separation of the peaks resolved in step 2, as well as enabling the PTH-derivatives of methionine, alanine/tryptophan, glycine, lysine, tyrosine, and threonine to be identified. In this segment of the development sequence, tryptophan appears as a shoulder on the alanine peak, but is not adequately resolved for identification purposes. The whole development sequence, during which all twenty protein PTH-amino acid derivatives can be separated (24), requires less than 1 h, only selected segments need be scanned to indicate individual amino acid derivatives, and standards can be run simultaneously with samples (up to 16 samples can be applied to a 10 x 10 cm plate and separated simultaneously) to improve sample identification from peak position locations.

One *disadvantage* of the multiple development technique concerns the separation of samples containing solutes of *similar polarity*. These tend to migrate together from the origin and become separated higher up the plate. For difficult separations fairly long plate lengths will be needed, and at each subsequent development sequence, a substantial amount of time is wasted while the solvent level reaches the level of the lowest spot on the plate. Under normal development conditions, the rate of solvent advance declines quadratically with distance above the solvent level. One solution to this problem is to move the position of solvent introduction to higher positions on the plate (25) or to remove a portion of the lower edge of the plate (26) for some or each subsequent development segment.

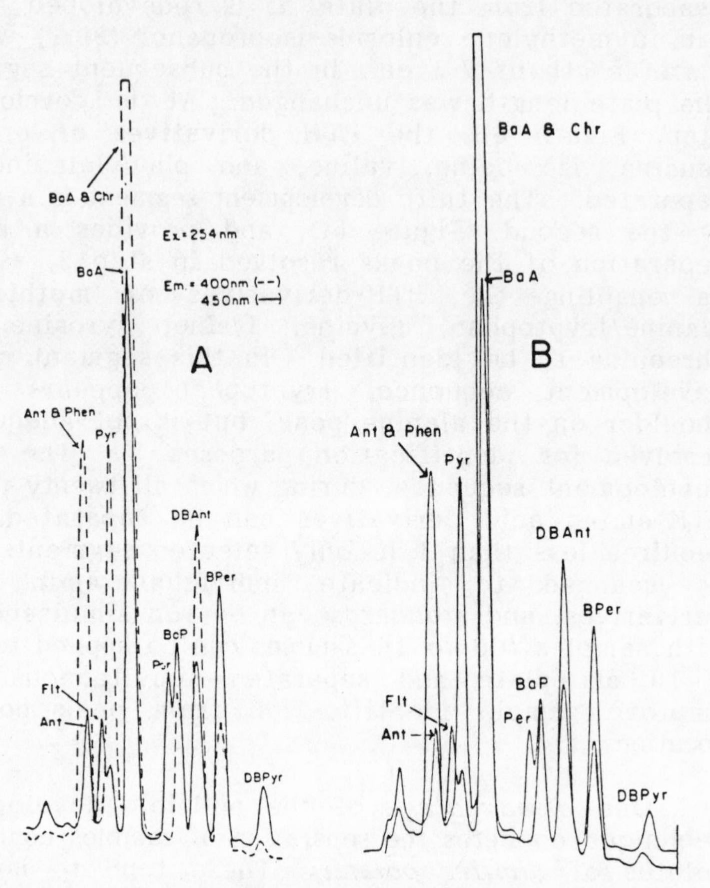

Figure 6. Separation of polycyclic aromatic hydrocarbons on octadecylsilanized silica gel HPTLC plates using the plate cutting technique. The mobile phase was methanol-acetonitrile-water (5:1:1). The plate was cut 0.5 cm from the lower edge before each successive development. The multiple development sequence was five 10-min., three 13-min., and two 15-min. developments. The separation obtained at the end of the fifth (A) and eleventh (B) development is shown.

Provided that the correct mobile phase velocity,

number of development segments, and time for each development are selected, then the plate cutting technique enables a spot to be made to traverse virtually the whole length of the plate without experiencing significant band broadening beyond that introduced in the first few developments. This is illustrated in Figure 6 for the separation of a mixture of polycyclic aromatic hydrocarbons scanned after five developments in (A) and ten developments in (B) (26). From the point of view of efficiency, by normal development an average value of ca. 2,000-3,000 theoretical plates was obtained; for normal multiple development this increased to ca. 5,000-10,000, while for multiple development with plate cutting at each segment, a value of 15,000-25,000 was obtained. In these experiments the mobile phase composition, stationary phase, and total development time were identical and ten segments were used in the multiple development experiments.

Two-Dimensional Sample Development

Unidimensional development, the method considered so far, is by far the most common development mode in TLC due to its speed of analysis and, furthermore, it is the only method used in practice for sample quantitation. Alternative development procedures are *two-dimensional* development, *circular* development, and *anticircular* development. In two dimensional TLC the plate is developed sequentially in two directions at right angles, the plate being dried between developments (27). If the same mobile phase is used for both developments, only a modest improvement in resolution can be anticipated corresponding to an approximate doubling of the number of theoretical plates. To obtain a useful improvement in sample resolution, it is necessary to arrange that the selectivity of the mobile phase (or sorbent if a bilayer plate is used) is different in both directions. A further problem is the near impossibility of obtaining quantitative data by scanning densitometry. The detection problem is presently being investigated by Guiochon using a photodiode strip along one edge of the plate and elution development as the second development to move the sample components across the detector (28). The detection and instrumental problems, however, remain formidable, and bidirectional column chromatography, the name given to the technique by Guiochon, is still under development.

Circular and Anticircular Developments

If the point of sample application and the point of entry of the mobile phase are at the center of the plate, then this mode of development is called circular chromatography (1).

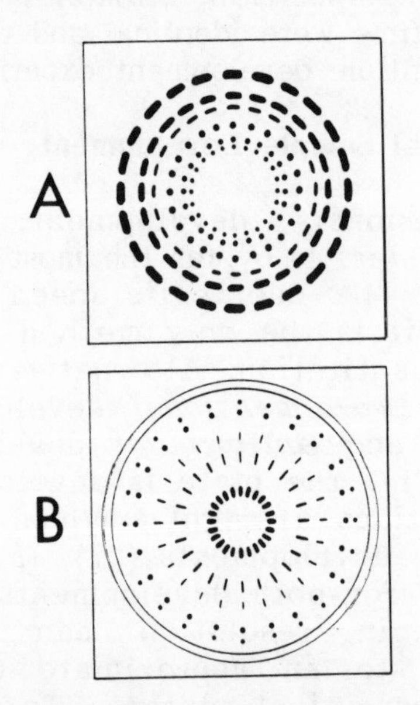

Figure 7. Circular development with the point of solvent entry at the plate center (A). Anticircular development from the outer circle towards the center (B).

Spots near the origin remain symmetrical and compact, while those near the solvent front are compressed in the direction of development and elongated at right angles to this direction, Figure 7A (3). In anticircular development the sample is applied along the circumference of an outer circle and developed toward the center of the plate (29). Spots near the origin remain compact, while those towards the solvent front are considerably elongated in the direction of migration, but changed very little in width when viewed at right angles to the direction of development, Figure 7B. The unique features of anticircular development are it's high speed and large sample capacity. For scanning individual chromatograms, densitometers must be capable of radial or peripheral scanning (2). Scanning in the direction of sample migration is termed radial scanning and the distance to the neighboring track assumes the dimensions of an angle in degrees. Scanning at right angles to the direction of development is termed peripheral scanning, analogous to the stylus tracking the groove in a record. Radial scanning is usually preferred over peripheral scanning.

SAMPLE APPLICATION

Samples are applied to TLC layers in the form of spots or bands to conform to the demands of minimum size and a homogeneous distribution of sample within the starting zone. Mass and volume overload of the layers has to be avoided if the separating potential of the layer is not to be compromised. For HPTLC plates the sample size should be reduced by a factor of 5- to 10-fold compared to conventional TLC layers. This means, for example, that sample volumes of 100-200 nL are typically used with spot applicators. Intuitively, one might anticipate a strong correlation between the dimensions of the starting zone and the subsequent size and resolution of the separated zones after development. This is true up to a point. Fenimore (2) has shown that the size of developed zones correlates more strongly with the quality of the layer than the size of the starting zone provided that the latter is maintained within reasonable limits. Spots of about 1.0 mm in diameter are optimum for modern HPTLC layers and there seems to be no advantage to reducing spot sizes below this dimension. Spots of significantly larger diameter will almost certainly compromise resolution and may generate an elliptical profile upon development.

Solvent Selection

The choice of sample solvent is also an important consideration in sample application (30). An ideal solvent should provide high sample solubility, be of low viscosity and sufficiently volatile to be easily evaporated from the plate. It must wet the sorbent layer, and it should be a weak chromatographic solvent for the sample on the sorbent selected for the analysis. These considerations cannot always be met in practice. Sorbents with chemically-bonded layers are poorly wetted by some solvents preventing penetration of the layer by the sample. This is particularly noticeable with reversed-phase layers and samples dissolved in aqueous organic solvent mixtures. In this case it may be necessary to use methanol, acetonitrile, methylene chloride, or acetone as the sample solvent and to minimize predevelopment of the spot by applying the sample repetitively in small volumes to the layer. As silica gel is wetted by nearly all solvents, the effect of predevelopment becomes the primary consideration for solvent selection. If possible, the most mobile sample component should have an R_F value not much greater than 0.1 if the sample solvent was used as a mobile phase for chromatographic development. A problem with such weak solvents is that they may not provide quantitative sample transfer from the application device which would dictate that a stronger solvent be used.

Sample Application Techniques

Samples may be applied as bands in two general ways; by using plates with *concentrating zones*, or by using a *band applicator*. HPTLC plates with concentrating zones are prepared from two layers of silica gel having different properties (31). The two layers abut each other, parallel to one edge, forming a very narrow interface. The concentrating zone is the narrower of the two zones, typically about 2 cm wide. The separating zone is prepared from the same silica gel used to coat normal HPTLC plates, while silica gel of large pore diameter and extremely low surface area (ca. 0.5 m^2/g) is used for the concentrating zone. Sample application is simplified because microliter volumes as spots or bands can be applied to any area of the concentrating zone. HPTLC plates with concentrating zones are particularly useful when large sample sizes are spotted, or crude samples (e.g., biological fluids) are applied directly to the plate. In the latter case, salts, biological polymers and other interferences that would bind with or change the homogeneity of the separation layer are retained in the concentrating zone and do not reach the analytical layer. During development, the sample migrates out of the concentrating zone and is focused at the interface as a narrow band. The sample band at the interface should be as narrow as possible, typically about 1 mm wide, and the sample should be distributed homogeneously throughout the band for quantitative analysis. A homogeneous sample band distribution may not be obtained unless the sample is carefully applied to the concentrating zone and, thus, HPTLC plates with concentrating zones are more frequently used in qualitative than quantitative analysis.

Band applicators work by *mechanically* moving the plate on a stage beneath a fixed sample syringe (2).

A controlled nitrogen atomizer sprays the sample from the syringe, forming narrow, homogeneous bands on the plate surface. The plate is moved back and forth under the spray nozzle to apply bands of any length between 0-199 mm and sample volumes between 2-99 µl. The rate of sample deposition is also controllable as mm/µL or s/µL. The advantage of band application is that for quantitative scanning densitometry the sample band can be made longer than the slit length of the light source which minimizes errors due to positioning of the sample within the light beam. The disadvantages are that relatively large sample volumes are required, the application process is slow, and if multiple samples of different concentrations are to be applied, then flushing the syringe between applications adds considerably to the time required for sample application.

For most quantitative work in HPTLC the sample is applied directly to the sorbent layer as a spot of small diameter. The most popular methods employ *fixed-volume dosimeters* comprising a platinum-iridium capillary of 100 or 200 nL volume sealed into a glass support capillary of larger bore (1). The capillary tip is polished to provide a smooth planar surface of small area (ca. 0.05 mm^2), which is brought into contact with the plate layer using a mechanical device to discharge its volume. It is important that the dosimeter does not deform the layer surface otherwise distorted sample spots may result on development. For this reason spotting by hand with a dosimeter is just about impossible. Mechanical application of the sample is made possible by attaching a metal collar to the glass support capillary so that it can be held by a magnet and lowered to the plate surface under controlled conditions. The simplest mechanical device is the "rocker" applicator, Figure 8. Here, an arm houses a magnet at one end to hold the dosimeter and a counter weight at the other to control the force with which the dosimeter strikes the plate surface. The dosimeter is both lowered to, and removed from, the plate's surface by a tipping action of the applicator arm about its fulcrum. A somewhat more sophisticated sample applicator is the Nanomat, Figure 8, which holds and lowers the dosimeter electromagnetically. The force with which the dosimeter engages the plate surface, the time it spends in contact with the layer, and the number of repetitions desired for complete sample transfer can be controlled

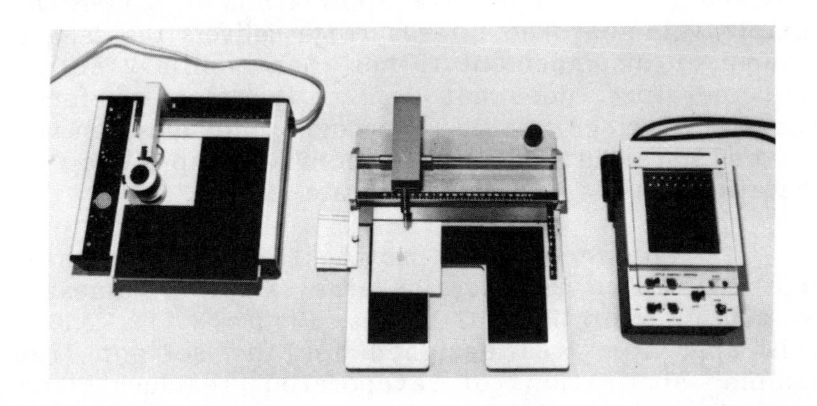

Figure 8. Sample applicators for HPTLC. From left to right, Nanomat, "rocker" applicator, and contact spotter.

electronically. Both applicators use a click-stop grid mechanism to aid in the even spacing of the samples on the plate and to provide a frame of reference for sample location during scanning densitometry. An automated sample spotting device has been developed which uses a flexible fused silica capillary tube as the applicator and a motor driven syringe to suck up and deposit sample volumes in the range of 100 nL to 20 µL (4). Controlled by a microprocessor, it can be programmed to select samples from a rack of vials and deposit fixed volumes of the sample, at a controlled rate, to selected positions on the plate. The applicator automatically rinses itself between sample applications and can spot a whole plate with different samples and standards without operator intervention.

As an alternative to the dosimeter, samples can be spotted with a *microsyringe* (2). For accurate dispension of nanoliter volumes, preselected in the range 50-230 nL, the syringe is controlled by a micrometer screw gauge. A fixed lever mechanism is provided for the repetitive application of a constant sample volume. The microsyringe delivers the sample volume by displacement rather than capillary action and, therefore, does not deform the plate surface. The microsyringe needle is brought only close enough to the plate surface for the convex sample drop of the ejected liquid to touch the plate surface.

The *contact spotter* provides a simple means of spotting viscous samples or large sample volumes up to 100 μL onto HPTLC plates, Figure 8 (12,32,33). This apparatus was designed for the solvent free sample application of evaporated residues from several samples simultaneously at precise locations on the thin-layer plate. The transfer medium is a fluorinated ethylene-propylene film coated with perfluorokerosene, positioned over a series of depressions in a metal platform. The film is forced to conform to the shape of the platform surface by applying slight vacuum through small holes in each depression. The sample solutions are pipetted into each depression and the solvent evaporated by gentle heat and a flow of nitrogen. The HPTLC plate is then positioned over the film, adsorbent side down, and with slight pressure replacing the vacuum, the spots are all transferred simultaneously to the plate. The various steps in the sample application process are summarized in Figure 9. Under conditions of low humidity, static charges may develop on the fluorocarbon film, causing movement of the samples out of the concave depressions on the surface of the spotter, which results in misalignment of the spots on the surface of the HPTLC plate. The static charge can be eliminated by wiping both surfaces of the film with antistatic paper moistened with ethanol before applying the samples or by using a variety of antistatic guns, radiation emitters, etc., available from record stores.

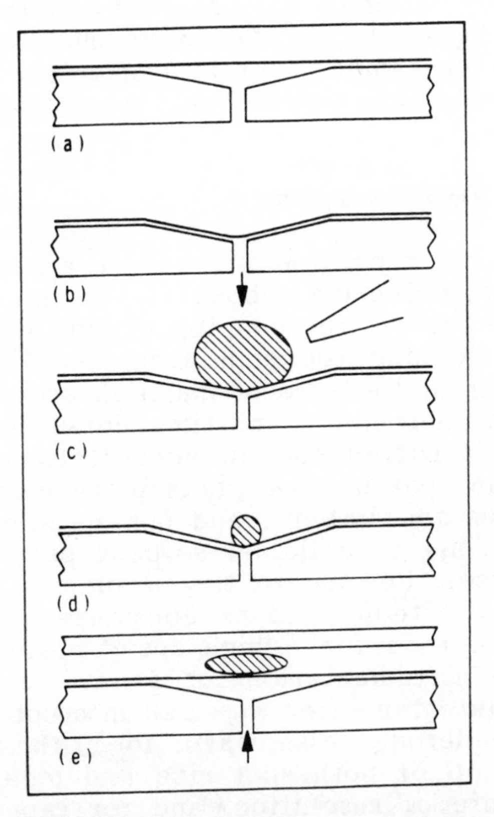

Figure 9. Sample application by contact spotting. A specially treated fluoropolymer film is pulled into a series of depressions in a metal plate by application of vacuum (a and b). Sample solution is delivered by pipette (c) and, after evaporation, a residue remains (d), which is transferred to the HPTLC plate by replacing the vacuum with slight pressure (e).

Crystalline samples do not penetrate the sorbent

layer, and they must be transferred in a small volume of a nonvolatile solvent such as octanol, dodecane, methyl myristate, acetophenone, etc. These carrier solvents may be added to the sample solution prior to evaporation to give a final volume of a few nanoliters, typically 5-20 nL, when the residue is transferred. A small amount of a colored dye that does not migrate in the chromatographic system can be added to each sample prior to evaporation to mark the exact position of the sample tracks for scanning densitometry.

SCANNING DENSITOMETRY

At best, inspection by eye of a TLC plate is capable of detecting about 1-10 µg of colored components with a reproducibility rarely better than 10-30%. Excising the separated spots, eluting the substance from the sorbent material, and measurement by solution photometry are time consuming and fairly inaccurate. Difficulties in accurately locating the edge of the spot by eye, incomplete elution of the sample from the sorbent, and non-specific background absorbance due to colloidal sorbent particles in the analytical solution add to the problem. The above process is too tedious to be considered an acceptable detection method for a high speed chromatographic technique. Instrumentation for *in situ* measurement of TLC chromatograms first appeared in about 1967 and is now considered essential for the accurate determination of both spot size and location, for a true measure of resolution, and for rapid, accurate quantitation.

Detection Methods

In situ measurements of substances on HPTLC plates can be made by a variety of methods: *reflectance, transmission, simultaneous reflectance* and *transmission, fluorescence quenching*, and *fluorescence* (2,5,6,30). Light striking the plate surface is both transmitted and diffusely scattered by the layer. Light striking a spot on the plate will undergo absorption so that the light, transmitted or reflected, is diminished in intensity at those wavelengths forming the absorption profile of the spot. The measurement of the signal diminution between the light transmitted or reflected by a blank zone of the plate and a zone containing sample provides the mechanism for quantitative measurements by absorption.

Absorption measurements may also be made by fluorescence quenching. In this case a special HPTLC plate incorporating a fluorescent indicator is used. When such a plate is exposed to ultraviolet (UV) light of short wavelength, the UV-absorbing spots appear dark against the brightly fluorescing background of lighter color. In this instance, the sample spots could be imagined as acting as UV filters: the spots absorb some of the excitation radiation, thus diminishing the intensity of the fluorescence emission eminating from the sample zones on the plate. Hence, they appear darker against a lighter background. Only those substances whose absorption spectra overlap the excitation spectra of the fluorescence indicator will be visualized by this process. Fluorescence quenching should primarily be considered as a visualization technique. It is less specific and sensitive than absorption measurements, in part due to severe background fluctuations resulting from the inhomogeneous distribution of fluorescent indicator in the sorbent layer.

The principal difference between absorption and fluorescence measurements is that, in the latter case, the illuminating radiation serves only to excite the examined material to emit secondary radiation at a longer wavelength than the exciting radiation. To a first approximation blank zones on the plate can be considered as "dark background" and the baseline, therefore, flat and free of noise generated by inhomogeneous scatter from the sorbent layer. Sample zones, therefore, behave as point sources of illumination superimposed on this "dark background". From the theoretical and practical point of view, this makes the problems of quantitation much easier as will be outlined later in the chapter.

Intrumentation for Scanning Densitometry

Commercial instruments for scanning densitometry share many features in common. Different lamps must be used as light sources in order to cover the entire UV-visible range from 200-800 nm. Halogen or tungsten lamps are used for the visible region and deuterium lamps for the UV region. High intensity mercury or xenon arc sources are preferred for fluorescence measurements. To select the measuring wavelength, either a monochromator or filter is used. Filter densitometers usually employ a mercury line source and filters to pass only light corresponding to individual wavelengths available from the source. Their advantage is their low cost, otherwise broad spectral sources and grating monochromators offer greater versatility for optimizing sample absorption wavelengths and are generally preferred. For fluorescence measurements, a monochromator or filter is used to select the excitation wavelength. A cutoff filter, which transmits the emission wavelength envelope but attenuates the excitation wavelength, is placed between the detector and the plate. Interference filters can be substituted for the cutoff filters if greater selectivity is required, albeit with some sacrifice in sensitivity. Photomultipliers or photodiodes are generally used for signal measurements.

Three optical geometries are predominantly used in contemporary scanning densitometers, Figure 10 (2, 5,6,30,34,35). The single-beam mode is the simplest optical arrangement and is capable of producing excellent quantitative results, but spurious background noise resulting from fluctuations in the source output, inhomogeneity in the distribution of extraneous absorbed impurities, and irregularities in the plate surface can be troublesome. These problems may be more severe for homemade coarse-particle TLC plates than for commercially prepared HPTLC plates.

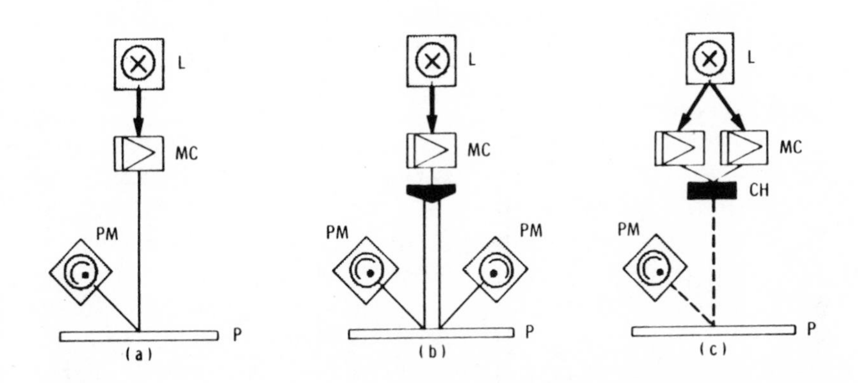

Figure 10. Schematic diagrams showing the optical arrangement of different kinds of scanning densitometers (a) single beam, (b) single wavelength double beam in space, and (c) dual wavelength single beam in time.

Background disturbances can be compensated for, to some extent, by double-beam operation. The two beams can either be separated in time at the same point on the plate, or separated in space and recorded simultaneously by two detectors.

The double-beam in space optical arrangement divides a single beam of monochromatic light into two beams that scan different positions on the plate. One beam scans the sample lane while the other traverses the blank region between sample lanes. The two beams are subsequently detected by matched photomultipliers and a difference signal is fed to the recorder; fluctuations in the source output are compensated in this way. As the two beams impinge on different areas of the plate, however, small irregularities in the plate surface and undesired background contributions from impurities in the sorbent layer may still pose problems.

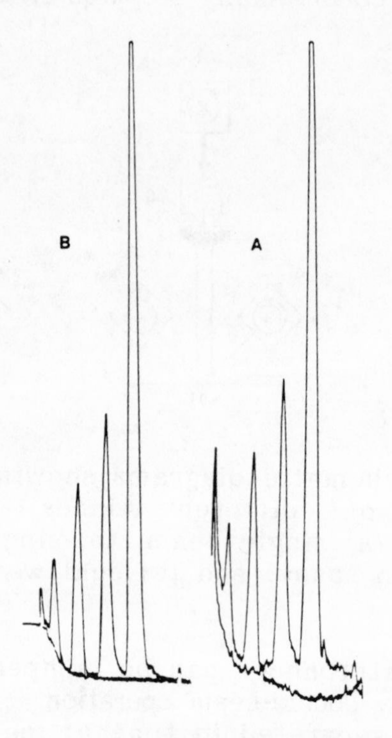

Figure 11. Use of background correction to improve baseline stability in the analysis of a mixture containing the drug metoprolol and some potential contaminants. Single-wavelength mode (a) λ = 280 nm and single-beam dual-wavelength mode (b) λ_1 = 280 nm and λ_2 = 300 nm. The background contribution resulting from spurious absorption by plate contaminants is entirely eliminated in (b).

In the single-beam dual-wavelength mode, fluctuations caused by scattering at a light absorbing wavelength (λ_1) are compensated for by subtracting the fluctuations at a different wavelength (λ_2) at which the spot exhibits no absorption but experiences the same scatter. The two beams are modified by a chopper and recombined into a single beam to provide the difference signal at the detector. As the scatter coefficient is to some extent wavelength dependent, the background correction is better when λ_1 and λ_2 are as nearly identical as possible. This requirement is often difficult to meet as absorption spectra are usually broad and two wavelengths at which absorption occurs in one and no absorption in the other may not be available. In favorable circumstances background correction in this mode can be very good as illustrated by Figure 11 (36).

In all contemporary densitometers, the position of the sample beam is fixed and the plate is scanned by mounting it on a movable stage controlled by stepping motors. For linear scanning the motor-driven stage transports the plate through the beam in the direction perpendicular to the slit. Each scan, therefore, represents a lane whose length is defined by the sample migration distance and whose width is determined by the slit dimensions. Some instruments can perform zig-zag, radial, and peripheral scanning as well.

Performance of a Scanning Densitometer

A question of some importance when assessing the performance of a scanning densitometer is how faithfully does it transform the separation on the plate into a strip chart chromatogram? Here the parameters of common interest are *component resolution*, *dynamic signal range*, and *sample detectability*, the latter being measured by the signal-to-noise ratio (5,17,30,37-40). Experimental variables can also affect these parameters, of which the most important are the slit dimensions governing the size of the measuring beam, the scan rate, and the total electronic time constant of the instrument and recording device. For perspective, the size of the measuring beam in a slit-scanning densitometer is defined by the slit width in the direction of scanning and the slit height in the orthogonal direction, as shown in Figure 12.

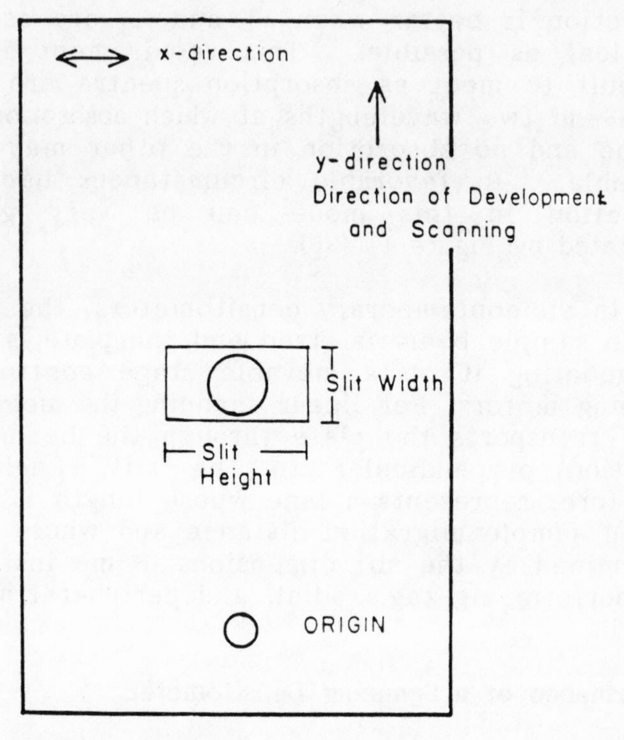

Figure 12. Enlargement of an HPTLC plate showing the orientation of the slit with respect to the direction of scanning.

No absolute method of ascertaining whether a densitometer can faithfully reproduce the chromatographic resolution from the plate is available. The spatial resolving power of a densitometer can be determined by scanning photographic test patterns or some similar standard (37,41). As these test patterns generally consist of narrow width equal-density squares or lines with discrete boundaries, they bear little resemblance to actual chromatographic separations. Commercially available densitometers have spatial resolving powers between 10 and 200 µm. The measurement of spatial resolving power provides information about the quality of the optical components of the densitometer and the extent of their misalignment, if any, but cannot be considered as an acceptable indicator of the ability of a densitometer to generate accurate chromatographic resolution data.

As a practical, although far from perfect solution to the above problem, the resolution of a partially separated pair of peaks can be compared to the values calculated from single standards run in separate neighboring lanes. In this way, experimental variables that cause a change in the perceived resolution can be discerned. This does not prove that the chromatographic resolution and that measured by the densitometer are identical, but does enable those operating conditions that result in a degradation of the chromatographic resolution to be identified.

Sensitivity

The sensitivity of a scanning densitometer is a complex function that depends on the quality of the electronic and optical components of the instrument. The influence of the measuring-beam dimensions on the recorded signal in the absorption mode is shown in Figure 13. The signal declines only slightly as the slit width is increased. A more dramatic change is observed with changes in the slit height. When the slit height is large compared to the diameter of the spot, the light flux emanating from blank regions of the plate is large compared to the contribution of absorbed light caused by the presence of the spot. The signal is thus weak. As the slit height is reduced to values close to the spot diameter, there is a substantial increase observed in the signal. This arises because the light flux transmitted/reflected by the blank area of the plate is diminished, while the amount of light absorbed by the spot under these circumstances remains constant.

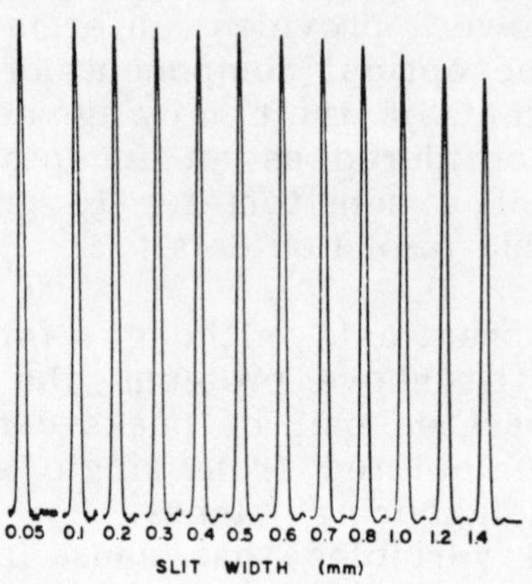

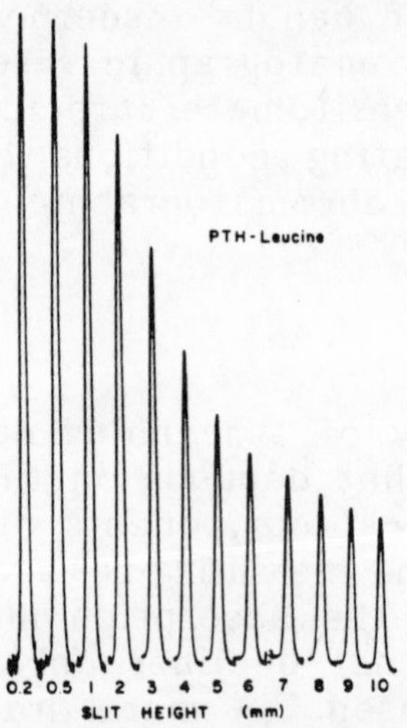

Figure 13. Variation of signal as a function of the slit dimensions for a slit-scanning densitometer operated in the absorption mode.

As the sample concentration across the diameter of the spot is not constant, the signal continues to increase as the slit height is reduced to less than the spot diameter.

The signal and signal-to-noise ratio do not necessarily follow the same trend. The signal-to-noise ratio is the more important parameter as it determines sample detectability.

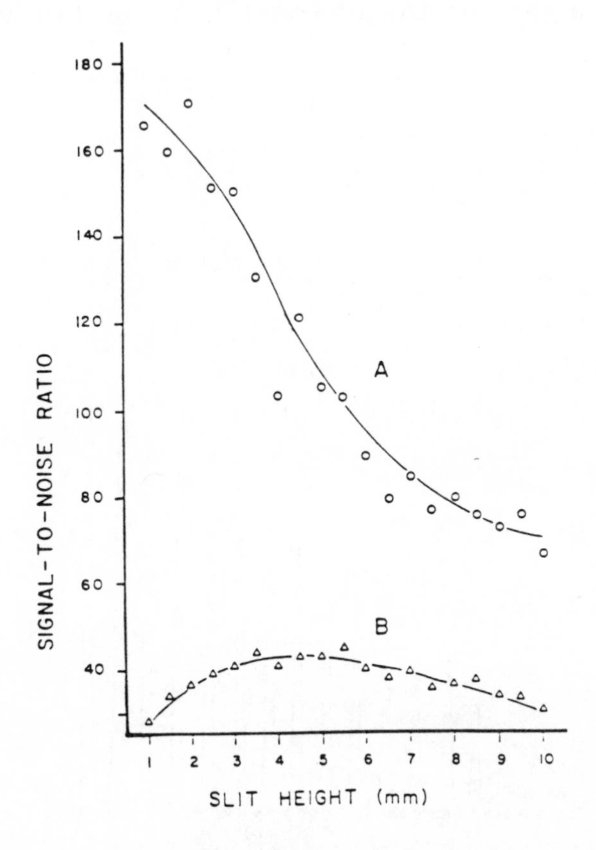

Figure 14. Relationship between signal-to-noise ratio and slit height at a constant slit width (0.5 mm) for (A) a compact spot, 3.0 mm, and (B) a diffuse spot, 6.6 mm, in slit-scanning densitometry.

Here the most important parameter is the *slit height* to *spot diameter ratio*; the influence of different slit width settings within their normal range is generally small by comparison. Figure 14 illustrates the change in signal-to-noise ratio, in the absorption mode, as a function of the slit height for two spots of different diameter. Both spots contain the same amount of sample. The maximum in the signal-to-noise ratio occurs at a different value of the slit height for the compact and the diffuse spot.

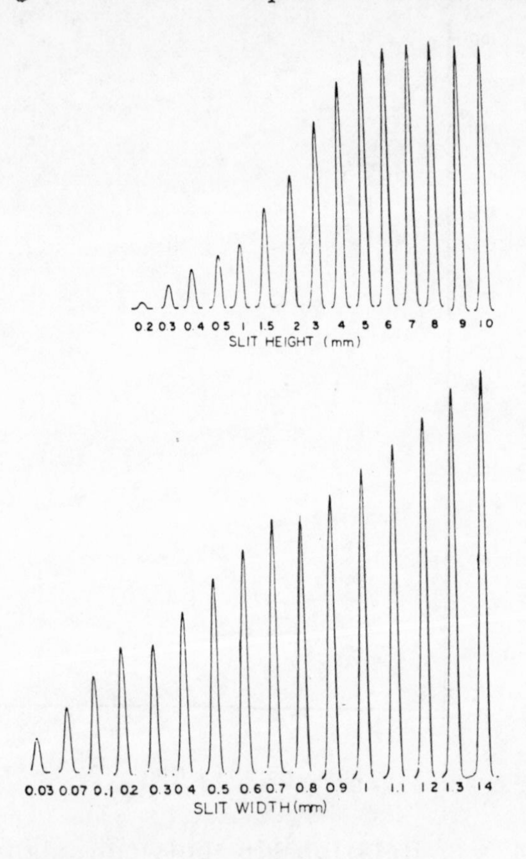

Figure 15. Variation of signal as a function of the slit dimensions for a slit-scanning densitometer operated in the fluorescence mode.

As in a normal separation spots of different sizes need to be scanned, it will not be possible to scan the complete separation under optimum sensitivity conditions. The maximum sensitivity is obtained when the slit height is smaller than the spot diameter. However, for practical reasons the use of small slit heights is not recommended, since if the sample track is not perfectly linear, then different zones of the spots will be scanned producing erroneous information. In this respect, there is a real advantage in using position scanning under computer control to accurately locate the center of each spot to be measured.

In the fluorescence mode, the influence of the measuring beam dimensions on sample detectability is not the same as that discussed for absorption measurements. The observed signal will depend on both the slit width and slit height dimensions, Figure 15. There is a regular decrease in signal in going from large to small slit width values. The reason being that, as the slit width is decreased, the amount of sample excited during the time the spot is being scanned, is reduced. Consequently, fewer molecules are fluorescing and the signal decreases. As far as the slit height is concerned, for values of the slit height less than the spot diameter, there is a regular increase in signal until the slit height reaches the same value as the spot diameter. At this point the signal intensity levels off and changes very little for larger slit height values. As can be seen from Figure 16, for spots of different diameter the signal maximum corresponds to a different slit height value. For clean plates, with little background fluorescence, there is only a small change in the noise signal as the area of the measuring beam is increased; the principal noise signal arising from lamp flicker and electronic noise. Thus, to maximize sample detectability for spots of different sizes, large slit height values are preferred; the minimum acceptable value being equivalent to the diameter of the largest spot to be scanned.

In the absorption and fluorescence mode there is a reciprocal relationship between signal and scan rate, Figure 17. Although slow scan rates lead to a larger accumulation of signal using electronic integrators, they also add to the time required to scan each track and in practice a compromise scan rate is usually selected.

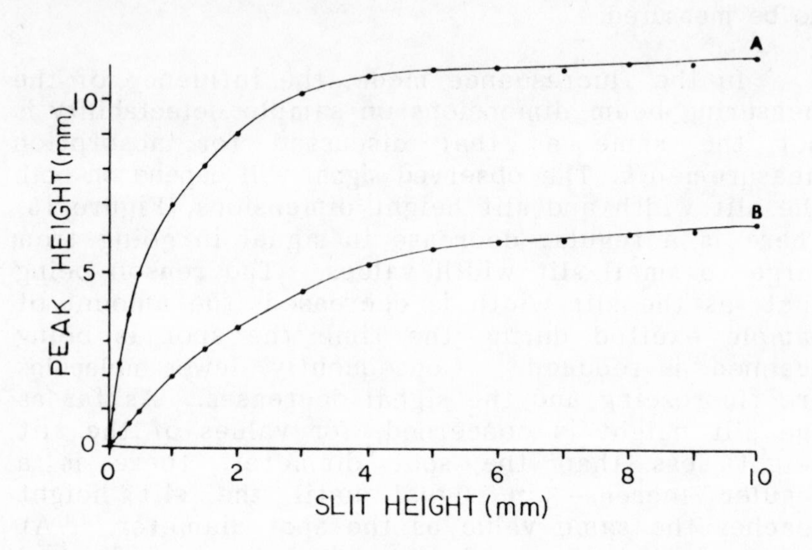

Figure 16. Variation of the fluorescence signal as a function of the peak height to the slit height. A = compact spot of 4.0 mm diameter and B = diffuse spot of 6.2 mm diameter.

Signal distortion will result at high scan rates if the electronic time constant of the instrument and recording device are too slow.

A protocol has been developed that provides a

practical and consistent method for comparing the performance of different slit-scanning densitometers as well as enabling detection limits and other sensitivity-dependent parameters to be determined under conditions of known sensitivity (17,42,43). Azobenzene and diphenylacetylene are suggested for use as standards.

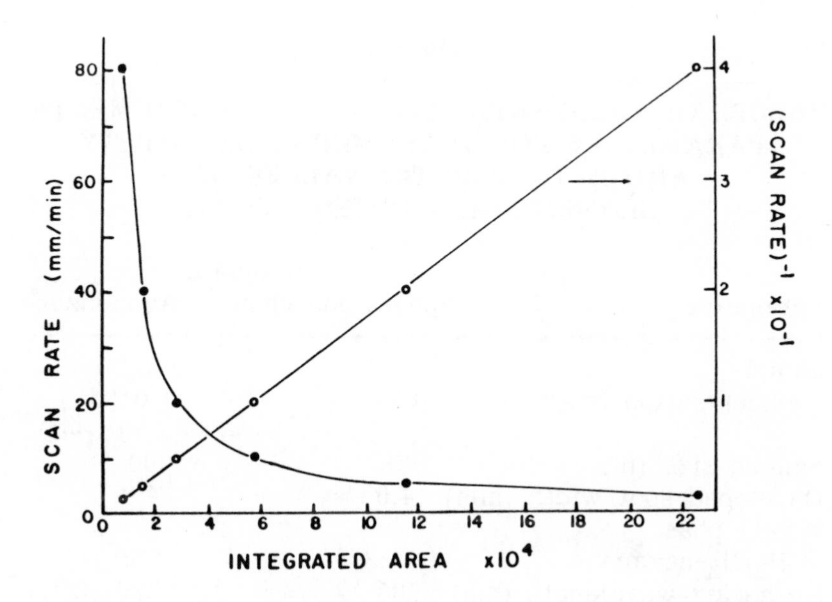

Figure 17. Relationship between scan rate and signal (as peak integrated area) in scanning densitometry.

Azobenzene is known to exist as cis and trans isomers which can be photolytically interconverted (44). The commercial material of m.p. 68°C is either the trans form or the photostationary equilibrium mixture of isomers. No change in the results obtained using azobenzene samples from several sources, or solutions stored in a refigerator for over one year has been noticed. Thus, azobenzene is sufficiently stable for use as a primary standard. It has two convenient absorption maxima for making measurement in either the UV or visible region and, being light yellow in color, it facilitates the alignment of the spot on the plate with the measuring beam. Diphenylacetylene is used as a secondary standard for measurements in the dual-wavelength single-beam mode. For this purpose a standard with a sharp UV cutoff is required. Chromatographic and instrumental parameters for determining sensitivity and detectability values

TABLE 1

VALUES OF CHROMATOGRAPHIC AND INSTRUMENTAL PARAMETERS FOR DETERMINING SENSITIVITY AND DETECTABILITY VALUES OF A SLIT-SCANNING DENSITOMETER

Parameter	Standard: Diphenylacetylene	Standard: Azobenzene
Sample concentration (mg/ml)	0.1	1.0 (UV) 5.0 (vis)
Sample size (nl)	200	200
Developed spot width (mm)	4.0	4.0
Mobile phase CH_2Cl_2-hexane	1:3	1:1
Measuring wavelength (nm)	290	320 (UV) 430 (vis)
Reference wavelength (Single-beam dual wavelength mode)	350	-
Scan mode	linear	linear
Scan rate (mm/min)	24	24
Slit width (mm)	0.5 & 0.8	0.5 & 0.8
Slit height (mm)	4.4	4.4
Recorder sensitivity (mV)	5 & 10	5 & 10
Recorder response (99% fsd,s)	0.33	0.33

are summarized in Table 1.

It should be noted that these conditions do not result in the maximum attainable signal but are justified because of their greater reliability and reproducibility of the measured data in routine use. Some typical data obtained for a commercially available slit-scanning densitometer is summarized in Table 2.

QUALITATIVE SAMPLE IDENTIFICATION BY SCANNING DENSITOMETRY

Most scanning densitometers make some provision for either manually or automatically recording the *in situ* spectra of any desired number of spots. For automatic spectrum recording, a motor-driven monochromator controlled by a central computer is used. For manual recording of spectra, the spot is scanned repetitively, while the monochromator position, or fluorescence emission filter for recording the fluorescence spectrum, is changed by fixed-wavelength increments between scans. A line connecting the individual peak maxima gives the substance-characteristic absorption or fluorescence emission envelope. The absorption spectra are rarely sufficiently characteristic for substance identification except by direct comparison with a standard measured on the same plate. The correspondence between solution spectra and *in situ* spectra can be quite poor in part due to the fact that the *in situ* spectra are usually measured under low resolution conditions. To maximize light throughput, most monochromators used in scanning densitometry have the band-pass in the range of 10 to 30 nm. There may also be some contribution to the measured spectra from plate absorption or scatter which may in itself change with the measuring wavelength. In the case of the fluorescence spectra, the solution and *in situ* spectra may show little correspondence due to the concert of energy loss and energy conversion

TABLE 2

SUMMARY OF SENSITIVITY AND DETECTABILITY DATA FOR THE CS–910 SCANNING DENSITOMETER

Standard	Mode	Wavelength (nm)	Signal (mm)	Noise (mm)	Sensitivity (mm/ng)	Detectability (ng)
Azobenzene	Transmission (single wavelength)	430	56.14 ± 1.81%	0.38	0.28 ± 1.8%	3.6 ± 1.85%
	Reflectance (single wavelength)	430	51.52 ± 1.58%	0.45	0.258 ± 1.58%	3.49 ± 1.61%
Diphenylacetylene	Reflectance (single wavelength)	290	66.0 ± 4.9%	3.0	3.27 ± 4.9%	1.84 ± 5.5%
	Reflectance (dual wavelength)	290	56.0 ± 7.2%	2.5	2.79 ± 7.2%	1.80 ± 8.1%

Sensitivity (S) = peak height/sample weight; detectability = 2 x noise/S. The noise signal and peak height are measured in mm for the baseline and scaled to the same recorder response setting. Sample weight is in ng.

mechanisms available to the sorbent sample.

It is less time consuming to scan a separation sequentially at several characteristic wavelengths than to record the full spectra of each spot (30,45). The ratios of the response values obtained at these characteristic wavelengths can be used to confirm the similarity between samples and standards or to indicate contamination of a sample spot with other components. The wavelengths selected for identification should emphasize the spectral characteristics of the substance of interest, for example, wavelengths corresponding to peaks and troughs in the spectra, etc. If standards are run on the same plate with the samples, then the reproducibility of absorbance response ratios is reasonable, RSD = 1-6% (46). Otherwise, the reproducibility of the absorbance response ratios will depend on how accurately the monochromator can be reset to a particular wavelength between measurements. For the same reason, the accuracy is improved if all samples and standards are first scanned at one wavelength and then the monochromator adjusted to the next position and the scanning repeated. Combining the information from coincidence of migration properties of samples and standards in the same chromatographic system and acceptable agreement between the absorbance or fluorescence emission response ratio is the most widely used technique for *in situ* substance identification in HPTLC.

Spectroscopic selectivity is much higher in the fluorescence mode because two different wavelengths, an excitation wavelength and an emission wavelength, are used for each measurement. The spectroscopic selectivity has been successfully utilized for qualitative sample identification (47,48). Additionally, the majority of organic compounds are not naturally fluorescent, but of those that are, several are of environmental or biological importance. In Figure 7, presented previously, the advantage of sequential wavelength scanning in the fluorescence mode was illustrated for a separation of a complex mixture of polycyclic aromatic hydrocarbons.

THEORETICAL CONSIDERATIONS FOR QUANTITATIVE ANALYSIS

All optical methods for the quantitative evaluation of thin-layer chromatograms are based upon measuring the difference in optical response between blank portions of the medium and regions where a separated substance is present. The propagation of light within an opaque medium is a very complex process that can only be solved mathematically if certain simplifying assumptions are made. The most generally accepted theory is that due to Kubelka and Munk (2,49). The Kubelka-Munk theory assumes that the transmitted and reflected components of the incident light are made up only of rays propagating inside the sorbent in a direction perpendicular to the plane of the plate surface. All other directions lead to much longer pathways and, therefore, much stronger absorption. Consequently, they contribute only negligibly to the total amount of tansmitted or reflected light. The restriction to propagation in only the forward and reverse direction does not apply to light exiting the medium, and at the plate/air boundary, light is distributed over all possible angles with the surface. By assuming that the scatter coefficient of the sorbent is unchanged by the presence of the sample, equation (5) can be derived.

$$\frac{(1 - R_{\infty})^2}{2R_{\infty}} = \frac{2.303}{S} \cdot a_m \cdot c \qquad (5)$$

where R_{∞} is the reflectance for an infinitely thick opaque layer,

a_m is the molar absorptivity of the sample,

c is the molar concentration of the sample
and S is the coefficient of scatter per unit thickness.

Although equation (5) relates the intensity of reflected light to sample concentration, it cannot be considered ideal for chromatographic purposes as it assumes a sorbent layer of infinite thickness. For a thin layer of thickness *Z*, the explicit hyperbolic solutions for reflectance and transmission, also derived by Kubelka and Munk, are more meaningful

$$I_R = \frac{\sinh [bSZ]}{a.\sinh[bSZ] + b.\cosh [bSZ]} \qquad (6)$$

$$I_T = \frac{b}{a.\sinh [bSZ] + b.\cosh [bSZ]} \qquad (7)$$

where I_R is the intensity of reflected light,
I_T is the intensity of transmitted light,
K_A is the coefficient of absorption per unit thickness,

$$a = \frac{SZ + K_A Z}{SZ}$$

and $b = (a^2-1)^{1/2}$

The application of equations (6) and (7) to quantitative analysis of thin-layer chromatograms is still quite complex as will be demonstrated later.

Many approaches can be used to calculate the fluorescence intensity of a sample in a thin ,opaque layer (35,50). The simplest phenomenological approach is based on the Beer-Lambert Law. If I_0 is the intensity of the incident beam and $\varnothing$ the quantum yield, then the fluorescence emission is given by equation (8)

$$F = \phi I_o(1 - e^{-a_m b_c c}) \quad (8)$$

where F is the fluorescence flux (emission),
ϕ is the quantum yield,
a_m is the molar absorptivity,
b_c is the thickness of the TLC layer
and c is the sample concentration.

For low sample concentrations the simplifying assumption, $e^{-a_m b_c c} = 1 - a_m b_c c$, can be made and equation (8) rearranged to give

$$F = \phi I_o a_m b_c c \quad (9)$$

As all terms in equation (9) are constant or fixed by the experiment, the fluorescence emission is linearly dependent on the sample amount. The fluorescence intensity is also independent of the spot shape provided that the spot is completely contained within the measuring beam. Experience supports the above equation in spite of the fact that it was derived without considering the influence of absorption and scatter by the medium on both the excitation and emission wavelengths. Similar results can also be derived from the Kubelka-Munk (50) equation which incorporates absorption and scatter by the chromatographic sorbent.

PRACTICAL CONSIDERATIONS FOR QUANTITATIVE ANALYSIS

Calibration Procedures in Absorption Mode

Reflectance and transmission measurements are particularly sensitive to small changes in experimental technique and materials. Variations in the layer thickness, layer quality (particle size and particle size distribution), spot shape and spot size, and uniformity of the sample development will affect the accuracy and precision of quantitative measurements. To improve precision, calibration standards should be run on each plate used for analysis. Calibration standards should be prepared in different concentrations and spotted as a fixed constant volume. Multiple spotting of a single standard solution to generate a calibration curve is not acceptable as this practice invariably results in the formation of a group of developed spots with different sizes. It should be noted that there is no simple correlation between signal response for a constant amount of substance and spot size (17).

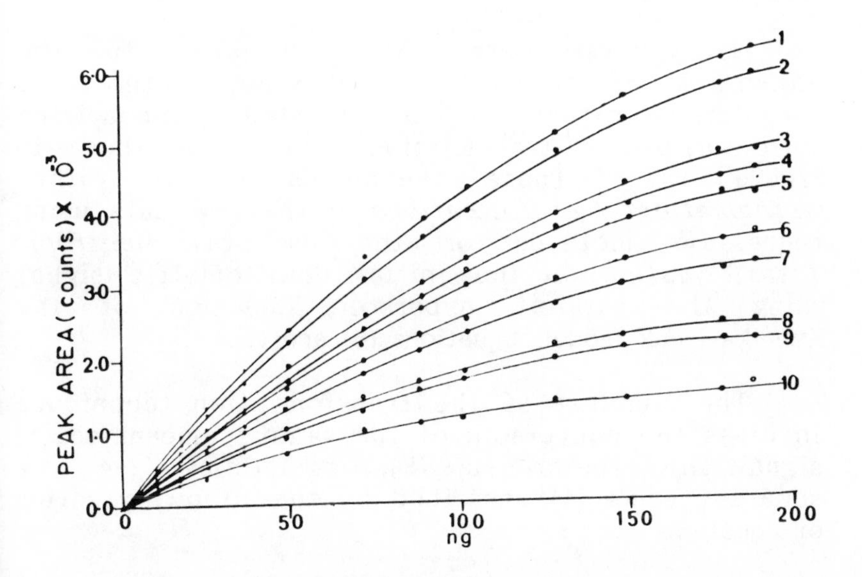

Figure 18. Calibration curves (peak area against sample size) in the absorption mode for some steroids. 1 = 17β-dihydroequilenin, 2 = 17α-equilenin, 3 = 17α-dihydroequilenin, 4 = 17β-estradiol, 5 = 17α-estradiol, 6 = equilin, 7 = 17β-dihydroequilin, 8 = estriol, 9 = estrone, and 10 = 17α-dihydroequilin.

Thus, standards and samples must have the same spot size for accurate quantitation.

Calibration curves for *in situ* reflectance and transmission measurement are inherently *non-linear*. This is entirely in agreement with the Kubelka-Munk theory. Some typical calibration curves are shown in Figure 18 (45). There is usually a narrow concentration range, over which the calibration curves can be considered as linear. The linear dynamic range of the calibration curve is substance-specific and, in favorable cases, may be adequate for the analysis.

Graphical Linearization Techniques

As analysts are less familiar with the properties of non-linear calibration methods, a considerable effort has been devoted to linearizing reflectance and transmission curves like those in Figure 18. These techniques involve either *mathematical transformations* of the raw data using regression methods or the use of *electronic transformation* of the initial densitometric signal using the explicit hyperbolic solution of the Kubelka-Munk model, equations (6) and (7).

The simplest of the transformation techniques involves the conversion of the sample amount and/or signal into reciprocals (2), logarithims (2-4), or squared terms (4) described by the following series of equations

$$\log R_e = a_o + a_1 \log m \qquad (10)$$

$$\frac{1}{R_e} = a_o + a_1 \left(\frac{1}{m} \right) \qquad (11)$$

$$\ln R_e = a_o + a_1 \ln m \qquad (12)$$

$$R_e^2 = a_o + a_1 m \qquad (13)$$

$$\ell n\ R_e = a_o + a_1 \sqrt{m} \tag{14}$$

where R_e is the reflectance (signal)
and m is the sample amount.

The use of equations (10), (11), and (13) to transform the calibration data for 17β-dihydroequilenin is shown in Figure 19. The extent of the linear range is shown by the solid line. In each case the region of the curve that is linearized is different; 50-100 ng for equation (10), 10-100 ng for equation (11), and 50-150 ng for equation (13).

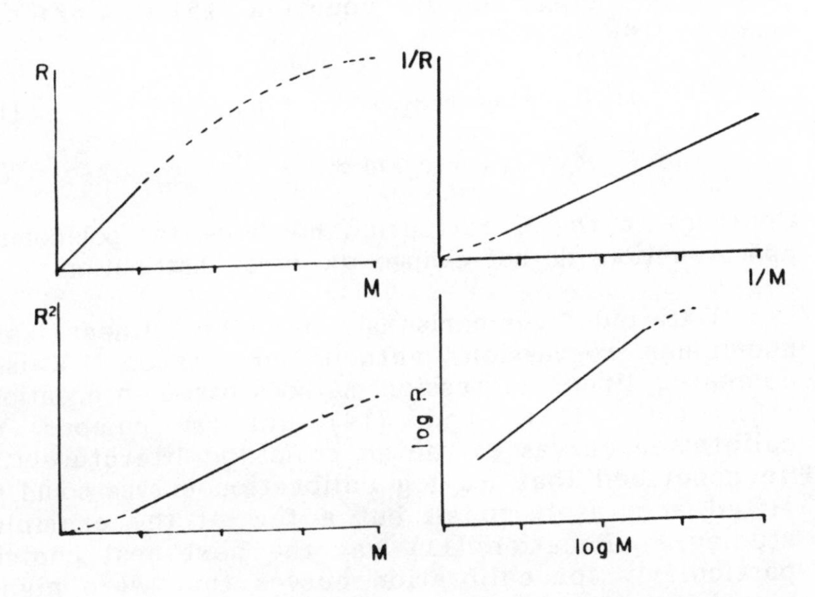

Figure 19. Linearization of the calibration curve for 17β-dihydroequilenin. All measurements were made by absorption in the reflectance mode and the calibration range in each case was 0-200 ng. The solid line represents the extent of the linear portion of the calibration curve. The broken line shows regions where a linear relationship does not exist.

No single method was successful for linearizing the complete calibration range.

The principal problem with the linearization methods is the manner in which errors become propagated (51,52). Errors in the original data are also transformed in the above methods leading to inhomogeneous variances in the transformed data and unreliable regression analysis.

One solution to the above problem is the use of non-linear regression analysis based on second-order polynomials described by equation (15) (2,4,53) and equation (16) (4).

$$\ell n\ R_e = a_o + a_1\ \ell n\ m + a_2(\ell n\ m)^2 \qquad (15)$$

$$R_e = a_o + a_1\ m + a_2\ m^2 \qquad (16)$$

Contrary to the linearization methods, the polynomial approximations do not change the error distribution.

Detailed comparisons of the linear and non-linear regression methods are rare. Kaiser compared linear regression methods based on equations (10), (11), (12), and (14) for a number of calibration curves extracted from the literature (2). He concluded that log-log calibration curves could be fitted accurately to all but a few of the examples studied. Equation (11) was the next best choice, particularly for calibration curves that were highly non-linear. Schmutz (4) compared equations (15) and (16) for the non-linear regression of three calibration curves. The best fit was obtained with equation (15). He also claimed that the fit was not compromised when as few as three standards were used. This is important from a practical standpoint as it minimizes the number of tracks utilized by calibration standards. De Spiegeleer et al. (53) compared six calibration methods for the determination of methyl nicotinate in pharmaceutical creams. The values obtained for the mean and standard deviations in the means for each of five calibration standards run on eight different plates are summarized in Table 3. The mean of the means was calculated from the differences between the calibration amounts applied to the plate and the calculated amounts from the regression equation, expressed as a percentage, averaged for each plate and then for all eight plates. The best fit was obtained with the second order polynomials, equations (15) and (16); of the linear approximations the reciprocal transformation, equation (11), gave the best fit.

TABLE 3

COMPARISON OF VARIOUS REGRESSION EQUATIONS FOR CALIBRATION IN TLC

Mathematical Model	Mean of Mean Value*	Mean of Standard Deviation*
Linear Equation	2.312	1.813
Equation (11)	1.310	1.340
Equation (12)	1.610	1.534
Equation (14)	1.843	1.519
Equation (15)	1.014	0.939
Equation (16)	1.074	1.102

*Expressed as % calculated from the difference between the calibration and calculated values.

Electronic Linearization Techniques

The Shimadzu 900 series of scanning densitometers has a built in curve linearizer for absorption measurements which utilizes the explicit hyperbolic solutions of the Kubelka-Munk theory (equations (6) and (7) given earlier) for the linear transformation. This series of densitometers also employs "flying spot" scanning and background correction to improve the accuracy of the transformation method (2,54). The sample track is scanned with a rectangular beam of small dimensions (0.4 x 0.4 mm for HPTLC) in a saw-tooth or square wave pattern by fixing the measuring beam in one position and mechanically moving the scanning stage in the x- and y-direction to generate the desired pattern. Zig-zag scanning, as it is called, has the advantage that the quantitative information obtained is independent of the spot shape and the distribution of the sample concentration within the spot. This will be true only if the beam size is small compared to the size of the spot and the measuring beam traverses across at least several zones of the spot. The signal produced during each swing of the measuring beam across the spot is converted to an absorbance value using equation (17)

$$D = \log (I_{R(o)}/I_{R(1)}) \qquad (17$$

where D is the observed response (absorbance),
$I_{R(1)}$ is the reflectance determined by equation (6)
and $I_{R(o)}$ is the reflectance determined for K_A = O.

It is further assumed that the sorbent layer does not absorb light and, therefore, the value of K_AZ represents the absorbance due to the sample. Electronic linearization of the signal is then possible by estimating the scatter component of the layer SZ. (Symbols, K_A, Z and S, are defined in equations (5), (6) and (7)). This can be done from plots of the kind shown in Figure 20 in which values for the absorbance D can be converted to values of K_AZ as a function of the scatter component SZ. Values for SZ are determined empirically for different sorbent media. Once a value of SZ is established, the value of K_AZ is obtained from the Kubelka-Munk equation and the total absorbance by integration of the values for each swing of the measuring beam through the spot. In the absence of the scatter term, the relationship between $K_A Z$ and sample concentration is linear, while the absorbance term (D) shows the expected curvature observed for normal calibration. To test the linearizer, several standards covering a reasonable calibration range should be scanned. A lack of linearity in the calibration curves indicates the wrong value of SZ, the scatter component, and a different value should be selected from the range of pre-selected values available.

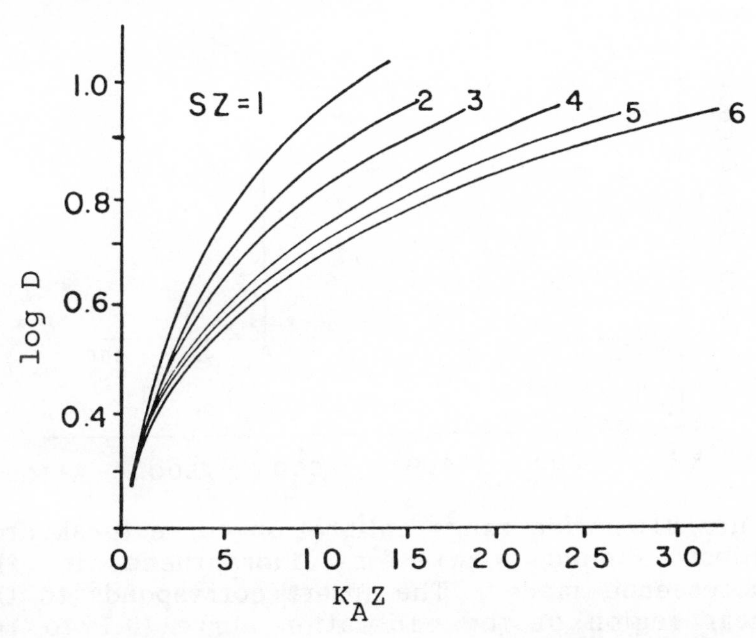

Figure 20. Relationship between the absorption signal (D) and sample amount ($K_A Z$) for different values of the plate scattering parameters (SZ).

Calibration Procedures in Fluorescence Mode

Calibration in fluorescence rarely produces problems. Calibration curves in fluorescence are usually linear over two to three orders of magnitude. An example is shown in Figure 21 for fluoranthene. The linear range for fluoranthene extends from the detection limit of 0.1 ng up to about 100 ng. For larger sample amounts it curves towards the height axis. For wide-range calibration the linearization techniques discussed for absorption could be applied here also, although this is not normally required.

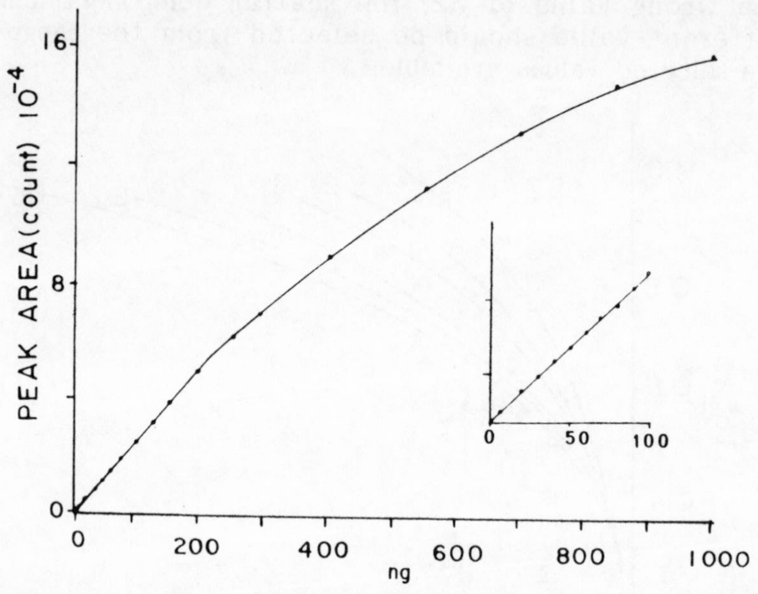

Figure 21. Wide range calibration curve (peak area against sample size) for fluoranthene in the fluorescence mode. The insert corresponds to the linear region of the calibration curve (0.1 to 100 ng).

A novel method of calibration for the routine screening of environmental extracts where multiple components have to be quantified has been proposed (48,55). Called the two-point calibration method, it has the advantage of requiring only a single standard for each substance to be determined and only a single track for all calibration standards. Thus, most of the plate is available for samples preserving the high sample throughput of the TLC method. It is based on the linear relationship observed between signal response and the dimensions of the slit width in fluorescence scanning densitometry (39)(as previously discussed).

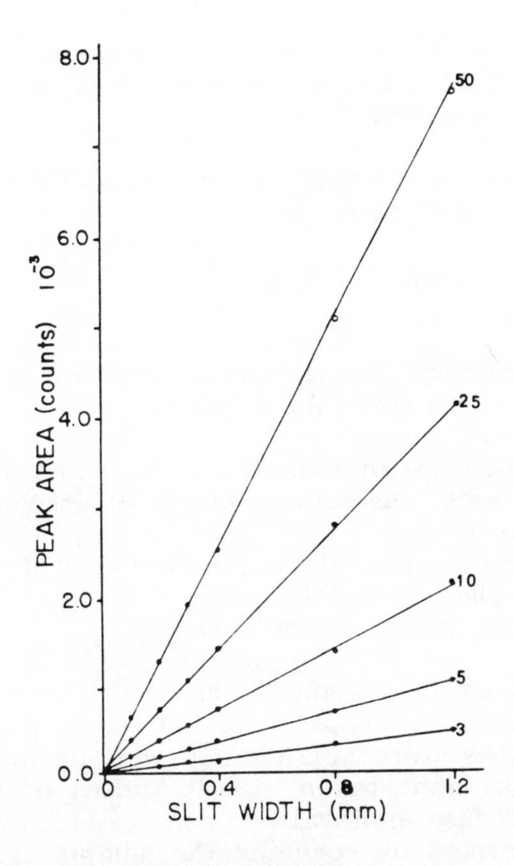

Figure 22. Relationship between signal (peak area) and slit width setting for various amounts of benzo[a]anthracene in the fluorescence mode using slit-scanning densitometry.

This relationship can be described by equation (18)

$$A_p = sW_w + b \qquad (18)$$

where A_p is the area of the peak,
s is the slope of the signal vs. slit width line,
W_w is the slit width
and b is the constant.

The slope of the detector response vs. slit width curve is proportional to the sample amount, as illustrated in Figure 22.

Thus, the relationship represented by equation (19) can be assumed to be valid.

$$m_u = m_s \left(\frac{s_u}{s_s} \right) \qquad (19)$$

where s_s is the slope of the detector response vs. slit width curve for a standard of known amount,
s_u is the slope of the detector response vs. slit width curve for a sample of unknown amount,
m_s is the weight of the standard applied to the plate
and m_u is the weight of the unknown.

The values for s_s and s_u can be determined by linear regression analysis. In practice, it was found that the slopes were sufficiently well defined by a single measurement taken at slit widths of 0.4 and 0.8 mm. Thus, calibration can be reliably and quickly performed by scanning the sample tracks and the track containing standards twice at two different slit widths. The agreement obtained between the calibration standards and calculated amounts for different sample sizes is shown in Table 4 (55).

TABLE 4

COMPARISON OF CALIBRATION STANDARDS AND CALCULATED AMOUNTS AT DIFFERENT SAMPLE SIZES (n = 15). THE REFERENCE STANDARD FOR THE TWO-POINT CALIBRATION WAS 10 ng

Calibration Standard (ng)	Calculated Amount by Two-Point Calibration (ng)
20.0	17.45 ± 1.30
15.0	13.97 ± 0.90
5.0	5.20 ± 0.30
2.0	2.22 ± 0.19
1.5	1.78 ± 0.14
1.0	1.14 ± 0.20
0.50	0.55 ± 0.09
0.20	0.24 ± 0.05
0.15	0.17 ± 0.05

The two-point calibration method cannot replace normal calibration when a high degree of accuracy is required. Its principal use is as a scouting method to determine approximate sample amounts.

Sources of Error in Scanning Densitometry

The main sources of error in scanning densitometry are the reproducibility of sample application (σ_V), reproducibility of the chromatographic conditions (σ_C), reproducibility of positioning the spot in the center of the measuring beam (σ_P), and reproducibility of the measurement (σ_m) (56). From the error propagation law, the total error (σ_t) is given by equation (20)

$$\sigma_t^2 = \sigma_V^2 + \sigma_C^2 + \sigma_P^2 + \sigma_m^2 \tag{20}$$

The error in the *measurement* (σ_m) can be determined by repeatedly scanning a single track of the TLC plate without changing any experimental variables between scans. It is composed of errors due to the optical measurement, electronic amplification, and the recording device. The value for σ_m is not constant for a particular instrument, it changes with respect to the signal-to-noise ratio, rising rapidly as the detection limit is approached. Steady values are obtained at sample amounts not obviously perturbed by baseline noise. For the value to be meaningful, the experimental conditions for the measurement should also be stated. For a properly adjusted instrument, σ_m values in the range 0.2-0.7% are typical (2,56).

The error of *positioning* the spot in the measuring beam is determined by repeatedly scanning a single track, as discussed above, except that in this case the first spot is repositioned in the measuring beam before each scan. The total error is then comprised of the positioning error and the measuring error (known from the previous experiment) and is evaluated from equation (21)

$$\sigma_p = (\sigma_t^2 - \sigma_m^2)^{1/2} \qquad (21)$$

The object is obviously to minimize this error which can become significant if many different plates have to be positioned during routine analysis. For manual positioning the use of slit height settings larger than the spot diameter will reduce positioning errors at the expense of some loss in sensitivity (as previously discussed).

The error due to sample *application* (σ_V) can be evaluated from scanning the first spot in the chromatogram which is assumed to be free of chromatographic error, that is, a spot of low R_F value. Again, by difference, the sample application error is given by equation (22)

$$\sigma_V = (\sigma_t^2 - \sigma_p^2 - \sigma_m^2)^{1/2} \qquad (22)$$

This error can be effectively eliminated by the use of an internal standard.

Having determined σ_m, σ_p, and σ_v, the *chromatographic error* can be determined by difference, equation (23)

$$\sigma_c = (\sigma_t^2 - \sigma_m^2 - \sigma_v^2 - \sigma_p^2)^{1/2} \qquad (23)$$

The chromatographic error may easily be the most significant error and is only reduced by minimizing the variability in the development process. The data pair technique (57) can be used to minimize errors due to migration differences as a result of edge effects, deviations in layer thickness, and non-linear solvent fronts, etc.

In modern scanning densitometry with HPTLC plates, the relative standard deviation from all errors can be maintained below 2%, making it a very reliable quantitative tool. The largest error will probably be the chromatographic error using modern sample application devices and scanning densitometers. Further improvements in the precision of HPTLC analysis will probably only be achieved by improvements to developing chambers and greater attention being paid to optimizing and standardizing the development process. It should be mentioned that the resolving power of the chromatographic system can be, in many instances, enhanced by using derivative recording. The latter provides a rapid, direct and simple means of verifying the homogeneity of any given spot so that the quantitative assay of severely overlapping components becomes a possibility (4,58,59).

THE USE OF COMPUTERS IN SCANNING DENSITOMETRY

Thin-layer chromatography is a data intensive technique. The operations of scanning, calibration, and quantitation can be both time consuming and tedious when performed manually. These procedures are also prone to variable errors such as those due to the positioning of the spot in the measuring beam and estimating the correct shape of the calibration curve. A further advantage of computers is that they simplify report writing by combining graphical, statistical, and word processing capabilities and simplify the archival process of data storage. Digital data stored on floppy disks can be reassessed and recalculated numerous times as the goals and interests of the analysis change without having to rerun the chromatogram. These and other considerations have been reviewed in the application of computers to thin-layer chromatography (2,4,34,60).

The computer can be used as an intelligent integrator performing all the calculations and smoothing of the digital data received from the densitometer without communicating with the densitometer in other respects (2,4,34,60,61). The link between the computer and the densitometer is made via an analog-to-digital converter. In more advanced instruments the densitometer is a slave to the computer which controls the scan function, selection of the measuring wavelength and other optional experimental variables (2,4,34,60). These variables are usually established in a pre-run dialog with the operator in accordance with a fixed menu of available options.

Automation of the scanning function is one of the more obvious parameters to control as, when this is achieved, the only operator intervention required is to position the plate in the densitometer and to remove it, at some preselected time later, when all the data has been recorded. In simple instruments a microprocessor is used for this purpose. Normally, the first track to be scanned is manually positioned in the measuring beam and then the values for track length, distance between tracks, and number of tracks to be scanned are entered. The densitometer then scans the plate following the established geometric pattern without further intervention by the operator. The disadvantage of this method is that if the samples migrate irregularly, then the spot may become misaligned with respect to the beam position and erroneous data generated. More sophisticated programs may recenter the measuring beam for the first spot of each track prior to scanning in a linear manner, execute a meander or zig-zag scan function, or optimize the beam spot co-ordinates for each spot in the chromatogram. The last of these options is called position scanning and is described in the literature (34).

Automatic scanning routines combined with video integration allow the correct determination of the baseline under each peak. During the scan mode, the background signal before and after each spot is acquired and used to compute the baseline position. The baseline in TLC may vary at different positions in the chromatogram due to matrix interferences and impurity gradients in the sorbent media. These can be eliminated from the signal by software control or with the intervention of the operator by video integration of the chromatographic data displayed on a monitor. Some progress has been made in subtracting the entire background contribution of the plate to the analytical signal with the intent of improving detection limits (4). With complete computer control of data acquisition and manipulation, the reproducibility of densitometric measurements is easily reduced to below 1% RSD. Thus, in current practice the densitometer is not the primary source of *variability* in quantitative TLC analysis.

PHYSICAL AND CHEMICAL METHODS USED TO ENHANCE SAMPLE DETECTABILITY

Visualization reactions were originally used in

thin-layer chromatography to enable colorless compounds to be detected by eye and, less frequently, to increase the selectivity of the detection process by reaction of the separated compounds with a reagent having chemical selectivity for a particular functional group or compound class. Several hundred reagents have been described for the above process (62-65). Many of these reactions were of a qualitative nature which was not a problem when quantitation of sample components was rarely performed by TLC. Some of these reactions have been adapted to the demands of quantitative scanning densitometry as either a *pre-* or *post-chromatographic treatment.* Post-chromatographic derivatization reactions for quantitative TLC have been reviewed by Ritter (4). The advantages of post-chromatographic methods are that by-products of the derivatization reaction do not interfere in the chromatography and all samples are derivatized simultaneously. The reagent is usually applied to the plate by dipping the plate into a dilute solution of the reagent or by spraying the reagent solution over the plate surface. For quantitative TLC, it is absolutely essential that the spraying is *uniform.* This is not easy to achieve by manual methods and motorized spray units are recommended (66). Dipping the plate into the reagent solution usually ensures an even application of reagent to the plate but is not without some difficulties. The reagent solvent must be a weak chromatographic solvent to avoid removal of the sample from the plate during dipping and zone broadening during the drying stage.

Pre-chromatographic derivatization reactions are usually favored when it is desired to modify the properties of the sample to enhance stability during measurement (i.e., minimize oxidative and catalytic degradation, etc.), to improve the extraction efficiency of the substance during sample cleanup, to improve the chromatographic resolution, or to simplify the optimization of the reaction conditions

(67). As both pre- and post-chromatographic methods *enhance* the sensitivity and selectivity of the detection process, a choice between the two methods will usually depend on the chemistry involved, ease of optimization, and which method best overcomes matrix and reagent interferences.

Fluorescence measurements are more selective and sensitive than absorption measurements. The formation of derivatives using the same techniques and reagents as used in liquid chromatography is very common (6).

From time to time a large disparity is observed between the fluorescence response of a substance in solution and the same amount of substance adsorbed onto a thin-layer plate. At least two general mechanisms are responsible for this: fluorescence quenching and catalytic decomposition. The extent of fluorescence quenching often depends on the sorbent medium and is frequently more severe for silica gel than for bonded-phase sorbents. The fluorescence emission spectra may also differ for the same substance in solution and on different adsorbents. Consider, for example, the fluorescence response of indeno[1,2,3-c,d]pyrene on a silica gel and an octadecylsilanized silica gel plate, Figure 23 (17). The response for identical amounts of the polycyclic aromatic hydrocarbon is significantly greater on the bonded-phase sorbent than for silica gel and the peak maxima in the fluorescence emission spectra are shifted to a lower wavelength as well. The mechanism or mechanisms of fluorescence quenching are not known with certainty, but it is generally assumed that adsorption onto silica gel provides additional nonradiative pathways for dissipation of the excitation energy that is at least partly relieved when silica gel is covered by bonded organic groups.

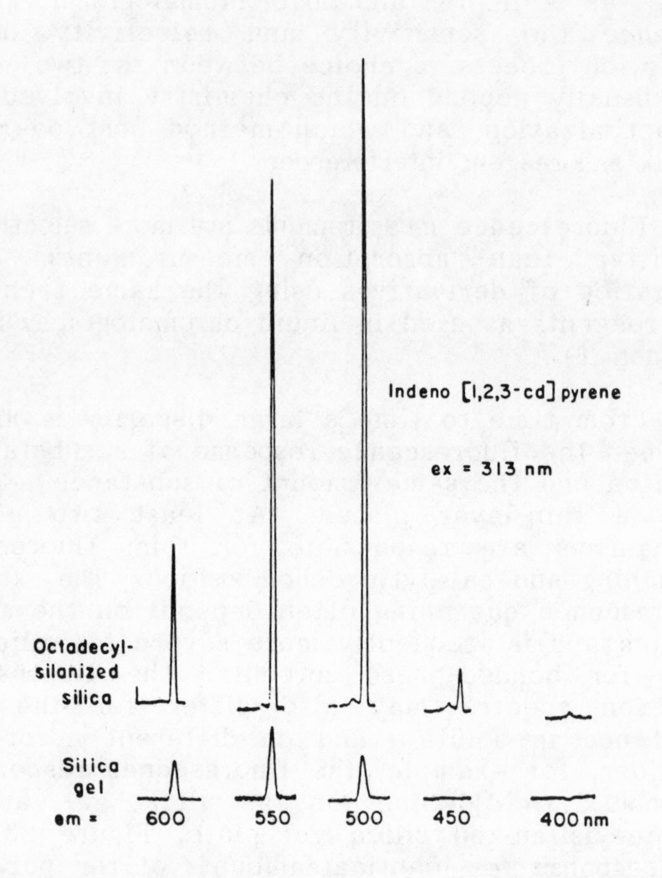

Figure 23. Change in fluorescence response for identical amounts of indeno[1,2,3-c,d]pyrene on silica gel and octadecylsilanized silica gel plates.

It has also been observed that dipping or spraying a developed plate with a solution of a viscous liquid such as liquid paraffin (68,69), glycerol (70), cyclodextrin (71), triethanolamine (72), Triton X-100 (17,68,70,73), or Fomblin H-Vac (17,73-75) prior to detection can enhance the fluorescence response of the sample, in favorable cases, by as much as 10- to

200-fold.

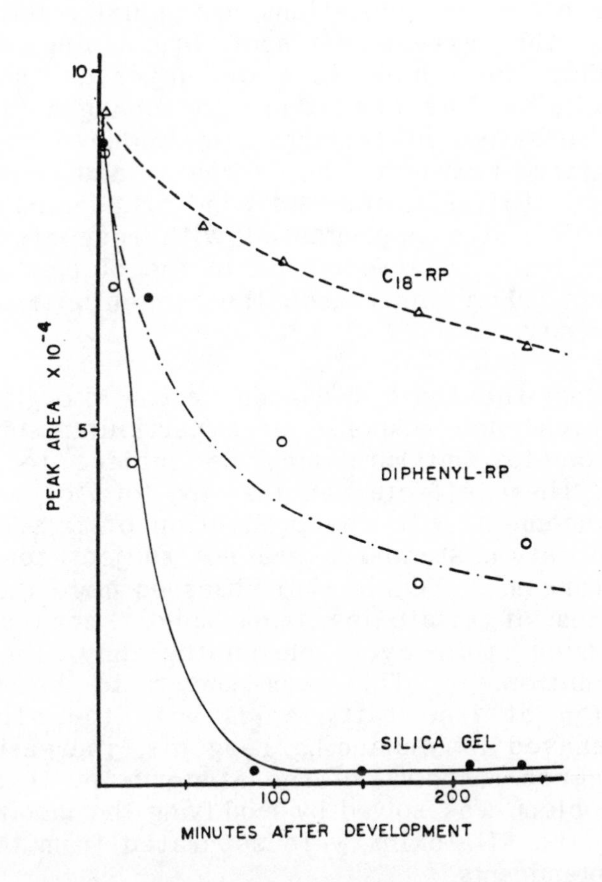

Figure 24. Rate of fluorescence decay of 1-aminopyrene (100 ng) on HPTLC plates coated with different sorbents. In the latter case the rate of fluorescence decay depends on the sorbent medium and is greatest for silica gel and lower for bonded-phase sorbents.

The fluorescence enhancement observed depends on a number of factors, of which the most important are the sample studied, the characteristics of the sorbent layer, the enhancement reagent used and its concentration, and the time between impregnating the plate with reagent and making measurements. In practice other considerations are equally important such as the extent of spot broadening due to maintaining the sample in a wet layer and shifts in the emission maxima due to changes in the sorbent/sorbate interactions or induced by new interactions between the sorbate and enhancing reagent. Diffusion-induced band broadening of the sample on plates impregnated with reagents of low viscosity leads to a reduction in the chromatographic resolution. For this reason the reagents used are of high viscosity.

From the above discussion we can also glean that the fluorescence response of a particular substance in a complex mixture may be subject to matrix effects. These effects can take the form of quenching or enhancement with the production of false data if the calibration standards are not subject to similar interferences. Zennie (76) observed more than 100% recoveries of aflatoxins from spiked corn samples when using thin-layer chromatography for their determination. This was shown to be due to co-elution of free fatty acids with the aflatoxins that behaved as enhancing reagents, increasing the fluorescence response of the aflatoxins by 14 to 36%. This problem was solved by modifying the mobile phase so that the aflatoxins were separated from the fatty acid contaminants.

Fluorescence quenching can also occur through chemical reaction while the sample is adsorbed on the sorbent. One general fluorescence quenching reagent is oxygen (77), although fortunately, its influence in most cases is quite small. In some cases autocatalytic oxidation can become a significant problem. Seifert (78) observed the oxidation of polycyclic aromatic hydrocarbons on silica gel plates to produce new products of low fluorescence yield.

In many cases these reactions show a time dependence that makes accurate quantitation difficult. This was observed to be the case for the fluorescamine derivatives of cephalosporins on silica gel plates (79) and 1-aminopyrene on different sorbents, Figure 24 (74,75). In instances of this kind impregnation of the plate with an antioxidant such as 2,6-di-tert-butyl-4-methylphenol (BHT) will diminish the rate of reaction substantially although it may fail to eliminate it entirely (68,74,75).

LESS COMMON METHODS OF SAMPLE IDENTIFICATION AND QUANTITATION

One weakness of thin-layer chromatographic methods is that they provide insufficient qualitative information to remove any doubt that a substance having a specific R_F value could not be mis-identified. This is a problem common to all chromatographic techniques and is usually solved by interfacing the chromatographic system to spectroscopic instruments. *Infrared spectra* of thin-layer chromatograms have been determined by *in situ* elution of individual spots with formation of potassium bromide micro pellets from the recovered residues (80), by *in situ* recording using diffuse reflectance FTIR (DRIFT) spectroscopy (81), and by photoacoustic spectroscopy after excising a portion of the sorbent containing the sample from the plate (82). The sensitivity of these methods is generally *inadequate* for modern TLC. Micrograms of sample are usually required to provide reasonable spectra and for functional group identification. The strong absorption by silica gel in the regions 3,700-3,100 and 1,650-800 cm^{-1} is a serious limitation for the *in situ* recording methods. *Mass spectra* of thin-layer chromatograms can be obtained by eluting sample spots from the sorbent or excising individual spots and introducing either the eluted residue or sample and sorbent directly into the ion source of the mass spectrometer (83), by fast atom bombardment (FAB) of excised spots (84), and by chemical ionization of sample spots thermally desorbed from TLC plates using either a low power pulsed carbon dioxide laser or a tungsten filament incadescent lamp (85). Microgram sample sizes are needed for FAB spectra, and although chemical ionization mass spectra could be obtained from nanogram amounts of volatile samples by the desorption techniques, the reproducibility was very poor. The interfacing of infrared and mass spectral methods to thin-layer chromatography, as currently practiced, are *lacking* in sensitivity, speed, and convenience to be used more than occasionally for special problems.

Two approaches have been made to develop *laser-based* scanning densitometers. Lasers can be used as sources for fluorescence measurements where their high power and small beam widths could be used to enhance sensitivity (86,87). However, to date, laser-based densitometers have not demonstrated any significant improvement in sample detectability over densitometers employing conventional arc sources.

Photothermal deflection, a development arising from studies on photoacoustic spectroscopy, has been evaluated as a detection technique for thin-layer chromatography (88). Detection limits have been obtained at least as good as those realized with conventional slit-scanning densitometers in the absorption mode.

A flame ionization detector has been also utilized in TLC. In this attempt, individual samples can be separated on thin rods, rather than plates, and the separated zones detected by moving the rods through a flame ionization detector (89,90). There are substantial differences of opinion as to the quantitative reliability of the technique (91,92).

Image analysis techniques can also be used for detection in TLC. In image analysis, a vidicon camera is used as the detector (93). Advantages of this approach are that it eliminates the need for mechanical scanning, it can process two dimensional chromatograms which are very difficult to quantify by conventional scanning densitometry, and it provides information very rapidly as the signal from all spots is accumulated simultaneously. The principal disadvantages of video densitometry at its present state of development are poor resolution and sensitivity compared to slit-scanning densitometers. The limit of detection in the absorption mode is not generally better than 0.1 μg for strongly absorbing substances. Resolution is limited by the number of pixels, picture elements, available with the vidicon.

For completeness the availability of specialized methods and equipment for measuring radioisotopes on thin-layer plates should be mentioned. The principal methods used are autoradiography, liquid scintillation counting, and direct scanning with radiation detectors (30,94).

SYNOPSIS

The introduction of fine-particle layer plates, new sample application and development devices, and further refinements in computer-controlled scanning densitometry have brought about a rebirth of interest in TLC. The performance of the technique has reached new heights by exploiting *forced-flow, continuous,* and *multiple development* techniques for sample separation. Under forced-flow conditions, the separation of the sample can be optimized *independently* of the plate length and mobile phase velocity which presently establish the upper bounds for separation using capillary-controlled flow systems. For capillary-flow controlled conditions, *short development lengths*, as commonly practiced in continuous and multiple development, are utilized to *maximize* the separating power of modern HPTLC plates by maintaining the separation within a reasonably optimized mobile phase velocity zone. To obtain *optimum performance* from HPTLC plates, it is necessary to ensure that *sample application*, either as spots or bands, is made in such a way as to provide *compact, homogeneous* starting zones. *Mechanical devices* for sample application are essential to achieve these goals and to provide mechanism for correctly positioning the sample zones in the probe beam of the scanning densitometer. Special *sample applicators* have been devised to provide automation or the ability to apply samples in the 1-100 µl range. After separation the sample zones are determined by *scanning densitometry* in the *reflectance* or *transmission* mode using *absorption* or *fluorescence*. Instrument parameters that influence *resolution* and *sensitivity* are the dimensions of the measuring beam controlled by the *slit dimensions*, the *scan speed*, and *electronic time constants* of the recording device. These are optimized for faithful recording of thin-layer chromatograms. The sensitivity is mainly determined by the ratio of the *slit height* to *spot diameter*. In the absorption mode, the *maximum* sensitivity is obtained when the slit height is *smaller* than the spot diameter, provided that the center of each spot is accurately located. In the fluorescence mode, the *maximum* sensitivity is achieved when *large* slit height values are utilized. *Calibration curves* are *nonlinear* in the absorption mode resulting in the use of several *mathematical linearization* or *electronic linearization* techniques. These techniques are applied very carefully as they may introduce substantial error. Fluorescence calibration curves are inherently *linear* up to limit imposed by self-absorption. The *sources* of *error* in scanning densitometry are the reproducibility of *sample application*, reproducibility of the *spot positioning* and reproducibility of the *measurement*. The most significant error is the *chromatographic error* that is reduced by minimizing the variability in the development process. The relative standard deviation from all errors can be maintained *below 2%*.

Computers are assuming an even greater role in scanning densitometry for automation of the densitometer and also to calculate and format all data associated with the measurement process. Physical and chemical methods are used to *enhance* sample *detectability* in those cases where the sample has a weak chromophore. The problem of fluorescence quenching observed for some sorbent/sample combinations is partially alleviated by application of novel *fluorescence enhancing reagents*. With the exception of optical methods, other detection techniques, such as *mass spectrometry, infrared spectroscopy, image analyzers* have yet to prove their worth when interfaced to TLC. *Rod-TLC* seems to be a promising technique but at present is limited by poor chromatographic performance and unreliable quantitation.

REFERENCES

1. "HPTLC: High Performance Thin-Layer Chromatography", A. Zlatkis and R. E. Kaiser (Eds.), Elsevier, Amsterdam, 1977.

2. "Instrumental HPTLC", W. Bertsch, S. Hara, R. E. Kaiser, and A. Zlatkis (Eds.), Huthig, Heidelberg, 1980.

3. D. C. Fenimore and C. M. Davis, Anal. Chem., 53 (1981) 253A.

4. "Proceedings of the Second International Symposium on Instrumental HPTLC", R. E. Kaiser (Ed.), Institute for Chromatography, Bad Durkheim,1982.

5. M. E. Coddens, H. T. Butler, S. A. Schuette and C. F. Poole, LC Magzn., 1 (1983) 282.

6. C. F. Poole and S. A. Schuette, "Contemporary Practice of Chromatography", Elsevier, Amsterdam, 1984.

7. H. Halpaap and J. Ripphahn, Chromatographia, 10 (1977) 613.

8. H. E. Hauck and H. Halpaap, Chromatographia, 13 (1980) 538.

9. W. Jost and H. E. Hauck, J. Chromatogr., 261 (1983) 235.

10. U. A. Th. Brinkman and G. de Vries, J. High Resolut. Chromatogr. Chromatogr. Commun., 5 (1982) 476.

11. U. A. Th. Brinkman and D. Kamminga, J. Chromatogr., 330 (1985) 375.

12. D. C. Fenimore and C. J. Meyer, J. Chromatogr., 186 (1979) 555.

13. H. Kalasz, Chromatographia, 18 (1984) 628.

14. H. F. Hauck and W. Jost, J. Chromatogr., 262 (1983) 113.

15. E. Mincsovics, E. Tyihak and H. Kalasz, J. Chromatogr., 191 (1980) 293.

16. R. E. Tecklenburg, G. H. Fricke and D. Nurok, J. Chromatogr., 290 (1984) 75.

17. C. F. Poole, M. E. Coddens, H. T. Butler, S. A. Schuette, S. S. J. Ho, S. Khatib, L. Piet and K. K. Brown, J. Liquid Chromatogr., 8 (1985) in press.

18. G. Guiochon and A. Siouffi, J. Chromatogr. Sci., 16 (1978) 470.

19. G. Guiochon and A. Siouffi, J. Chromatogr. Sci., 16 (1978) 598.

20. A. Siouffi, F. Bressolle, and G. Guiochon, J. Chromatogr., 209 (1981) 129.

21. D. Nurok, R. M. Becker, and K. A. Sassic, Anal. Chem., 54 (1982) 1955.

22. K. Burger, Fresenius Z. Anal. Chem., 318 (1984) 228.

23. C. F. Poole, Trends Anal.Chem., 4 (1985) 209.

24. S. A. Schuette and C. F. Poole, J. Chromatogr., 239 (1982) 251.

25. D. Nurok, Anal. Chem., 53 (1981) 714.

26. C. F. Poole, H. T. Butler, M. E. Coddens, S. Khatib and R. Vandervennet, J. Chromatogr., 302 (1984) 149.

27. M. Zakaria, M.-F. Gonnord and G.Guiochon, J. Chromatogr., 271 (1983) 127.

28. G. Guiochon,M.F. Gonnord, M. Zakaria, L. A. Beaver and A. M. Siouffi, Chromatographia, 17 (1983) 121.

29. R. E. Kaiser, J. High Resolut. Chromatogr. Chromatogr. Commun., 1 (1978) 164.

30. C. F. Poole, H. T. Butler, M. E. Coddens and S. A. Schuette, in "Analytical and Chromatographic Techniques in Radiopharmaceutical Chemistry", D.M. Wieland, T. J.Manger and M. C. Tobes (Eds.), Springer-Verlag, New York, 1985, p. 3.

31. H. Halpaap and K.-F. Krebs, J. Chromatogr., 142 (1977) 823.

32. G. Malikin, S. Lam and A. Karmen, Chromatographia, 18 (1984) 253.

33. C. M. Davis and C. A. Harrington, J. Chromatogr. Sci., 22 (1984) 71.

34. I. M. Bohrer, Topics in Current Chem., 126 (1984) 95.

35. R. J. Hurtubise, "Solid Surface Luminescence Analysis", Dekker, New York, 1981.

36. M.-L. Cheng and C. F. Poole, J. Chromatogr., 257 (1983) 140.

37. H. T. Butler, S. A. Schuette, F. Pacholec and C. F. Poole, J. Chromatogr., 261 (1983) 55.

38. H. T. Butler, F. Pacholec and C. F. Poole, J. High Resolut. Chromatogr. Chromatogr. Commun., 5 (1982) 580.

39. H. T. Butler and C. F. Poole, J. High Resolut. Chromatogr. Chromatogr. Commun., 6 (1983) 77.

40. M. E. Coddens, S. Khatib, H. T. Butler and C.F. Poole, J. Chromatogr., 280 (1983) 15.

41. R. Apothekar, in "Advances in Thin-Layer Chromatography: Clinical and Environmental Applications", J. C. Touchstone (Ed.), Wiley, New York, 1982, p. 149.

42. M. E. Coddens and C. F. Poole, Anal. Chem., 55 (1983) 2429.

43. M. E. Coddens and C. F. Poole, LC Magzn., 2 (1984) 34.

44. G. Zimmerman, L.-Y. Chow and U.-J. Paik, J. Amer. Chem. Soc., 80 (1958) 3528.

45. L. Zhou, C. F. Poole, J. Triska and A. Zlatkis, J. High Resolut. Chromatogr. Chromatogr. Commun., 3 (1980) 440.

46. K. Y. Lee, C. F. Poole and A. Zlatkis, Anal. Chem., 52 (1980) 837.

47. H. T. Butler, M. E. Coddens and C. F. Poole, J. Chromatogr., 290 (1984) 113.

48. H. T. Butler, M. E. Coddens, S. Khatib and C. F. Poole, J. Chromatogr. Sci., 23 (1985) 200.

49. V. Pollak, Adv. Chromatogr., 17 (1979) 1.

50. M. Prosek, A. Medja, E. Kucan, M. Katic and M. Bano, J. High Resolut. Chromatogr. Chromatogr. Commun., 3 (1980) 183.

51. S. Ebel, Topics in Current Chem., 126 (1984) 71.

52. S. Ebel, D. Alert and U. Schaefer, Chromatographia, 18 (1984) 23.

53. B. De Spiegeleer, W. Van den Bossche and P. De Moerloose, Chromatographia, 20 (1985) 249.

54. H. Yamamoto, T. Kurita, J. Suzuki, R. Hira, K. Nakano, H. Makabe and, K. Shibata, J. Chromatogr., 116 (1976) 29.

55. H. T. Butler and C. F. Poole, J. Chromatogr. Sci., 21 (1983) 385.

56. S. Ebel and E. Glaser, J. High Resolut. Chromatogr. Chromatogr. Commun., 2 (1979) 36.

57. H. Bethke, W. Santi and R. W. Frei, J. Chromatogr. Sci., 12 (1974) 392.

58. V. Such, J. Traveset, R. Gonzalo and E. Gelpi, J. Chromatogr., 234 (1982) 77.

59. J. Traveset, V. Such, R. Gonzalo and E. Gelpi, J. High Resolut. Chromatogr. Chromatogr. Commun., 5 (1982) 483.

60. V. Pollak, J. Liquid Chromatogr., 3 (1980) 1881.

61. S. Ebel, Fresenius Z. Anal. Chem., 318 (1984) 201.

62. J. G. Kirschner, "Thin-Layer Chromatography", Wiley, New York, 2nd ed., 1978.

63. J. C. Touchstone and M. F. Dobbins, "Practice of Thin-Layer Chromatography", Wiley, New York, 1978.

64. E. Stahl, "Thin-Layer Chromatography", Springer-Verlag, New York, 1969.

65. G. D. Barrett, Adv. Chromatogr., 11 (1974) 145.

66. F. Kreuzig, Chromatographia, 13 (1980) 238.

67. W. Funk, Fresenius Z. Anal. Chem., 318 (1984) 206.

68. D. Wollbeck, E. v. Kleist, I. E. Elmadfa and W. Funk, J. High Resolut. Chromatogr. Chromatogr. Commun., 7 (1984) 473.

69. W. Funk, R. Kerler, E. Boll and V. Dammann, J. Chromatogr., 217 (1981) 349.

70. S. Uchiyama and M. Uchiyama, J. Liquid Chromatogr., 3 (1980) 681.

71. A. Alak, E. Heilweil, W. L. Hinze, H. Oh and D. W. Armstrong, J. Liquid Chromatogr., 7 (1984) 1273.

72. A. Oztunc, Analyst, 107 (1982) 585.

73. S. S. J. Ho, H. T. Butler and C. F. Poole, J. Chromatogr., 281 (1983) 330.

74. K. K. Brown and C. F. Poole, LC Magzn., 2 (1984) 526.

75. K. K. Brown and C. F. Poole, J. High Resolut. Chromatogr. Chromatogr. Commun., 7 (1984) 520.

76. T. M. Zennie, J. Liquid Chromatogr., 7 (1984) 1383.

77. F. De Croo, G. A. Bens and P. De Moerloose, J. High Resolut. Chromatogr. Chromatogr. Commun., 3 (1980) 423.

78. B. Seifert, J. Chromatogr., 131 (1977) 417.

79. H. Fabre, M.-D. Blanchin, D. Lerner and B. Mandrau, Analyst, 110 (1985) 775.

80. H. J. Issaq, J. Liquid Chromatogr., 6 (1983) 1213.

81. G. E. Zuber, R. J. Warren, P. P. Begosh and E. L. O'Donnell, Anal. Chem., 56 (1984) 2935.

82. R. L. White, Anal. Chem., 57(1985) 1819.

83. J. Henion, G. A. Maylin and B. A. Thompson, J. Chromatogr., 271(1983) 107.

84. T. T. Chang, J. O. Lay and R. J. Francel, Anal. Chem., 56 (1984) 109.

85. L. Ramaley, M.-A. Vaughan and W. D. Jamieson, Anal. Chem., 57 (1985) 353.

86. M. K. L. Bicking, R. N. Kniseley and H. J. Svec, Anal. Chem., 55 (1983) 200.

87. P. B. Huff and M. J. Sepaniak, Anal. Chem., 55 (1983) 1992.

88. K. Peck, F. K. Fotiou and M. D. Morris, Anal. Chem., 57 (1985) 1359.

89. R. G. Ackman, Methods Enzymol., 72 (1981) 205.

90. H. O. Ranger, J. Liquid Chromatogr.,4 (1981) 2175.

91. T. N. B. Kaimal and N. C. Shantha, J. Chromatogr., 288 (1984) 177.

92. A. K. Banerjee, W. M. N. Ratnayake and R.G. Ackman, J. Chromatogr., 319 (1985) 215.

93. S. Pongor, J. Liquid Chromatogr., 5 (1982) 1583.

94. S. D. Shulman, J. Liquid Chromatogr., 6 (1983) 35.

Quantitative Analysis using
Chromatographic Techniques
Edited by Elena Katz

Chapter 7

CHROMATOGRAPHY AS A QUANTITATIVE TOOL IN PHARMACEUTICAL ANALYSIS

Eric C. Jensen

INTRODUCTION

Chromatography methods have been widely adapted in the pharmaceutical field for many of the same reasons that they are used in other areas. The selectivity, sensitivity, and convenience of chromatography make the technique a powerful tool which permits the analyst to make precise and accurate determinations of drugs, their related impurities and degradation products in a variety of sample matrices. The goal of this chapter is to review how and where chromatographic methods fit into the sample scheme encountered in pharmaceutical products. The details of the chromatographic process have been discussed elsewhere in this text and will not be considered here.

The topics which are discussed will include separation development, sample handling, factors which determine the precision and accuracy of quantitative analyses, and examples of the chromatographic methods used in some pharmaceutical applications. The various sample types commonly encountered will be examined in light of what chromatographic methods are expected to do and how the samples are handled prior to the chromatographic experiment. A complete discussion of the use of quantitative chromatographic methods for pharmaceutical applications is not possible in this chapter. Rather, the goal will be to cover the more important aspects of method development for pharmaceutical samples. Many of the method development and validation topics discussed here are applicable to sample types other than pharmaceuticals. The main focus of the discussion will center on analytes which come from the pharmaceutical field and on the aspects of chromatographic method development important to this class of compounds.

SEPARATION DEVELOPMENT

The chromatographer in the pharmaceutical field must deal with a variety of challenges. The samples that one confronts are such that the chromatographer needs to understand the nature of the samples, so that all the relevant components may be accounted for. There exists both a moral and legal obligation that pharmaceutical products intended for human consumption be characterized as *completely* as possible. Quantitative chromatographic methods are a powerful tool for elucidating this type of information. These methods are also used for the quantitation of pharmaceutical products and other biologically active compounds in biological matrices. In order to discuss the methods by which a chromatographer develops these separations, it is first important to examine the types of samples which require these separation methods. In the pharmaceutical industry, the primary sample types encountered by the chromatographer include bulk drug substances, drug products, starting materials and intermediates used in the preparation of the drug, and biological samples which contain small amounts of the parent drug as well as metabolites. In each situation the type of information sought by the chromatographer and the approach taken toward method development is different.

It should be noted that the principles discussed here apply to HPLC, GC and TLC equally well. For the evaluation of the bulk drug substance, the drug itself is the primary component of the sample. Chromatographic methods for this material are required to do two things. First, they must provide a quick, reliable determination of the purity of the drug by using an assay that employs a well-characterized reference standard. This particular chromatographic experiment may be the easiest to conduct as the components are often of a very high purity, and little or no sample preparation, beyond solubilization of the analyte, is required. However, there is more to this type of quantitative determination than it appears to be. The second requirement for this method is that all other components of the sample should be well separated from the peak for the main component. This challenge becomes significant when the chromatographer is faced with developing a method which, not only provides acceptable chromatography for the drug, but which also is capable of separating its synthetic precursors, side-reaction products, and all potential degradation products. These components may have unwanted pharmacological or toxicological effects which can outweigh any benefit from the administration of the drug. The determination of these impurities, using chromatographic methods, provides a tool for their control and for the protection of the patient who ultimately receives the drug. Depending on the efficiency of the final cleanup process used for the material, and on the purity of the various starting materials and intermediates, there are a number of possible ways to explain the presence of related substances in the sample. Consultation with the organic chemists involved with the synthesis of the compound can often lend valuable insight as to the types of impurities that may be present. Often the purity of the final product may be aided by controlling the purity of the materials used in the synthesis of the final product. This is particularly important when geometrical or optical isomers may be present in one or more steps of the synthesis. The use of chromatographic methods for the control of the starting materials and the intermediates is an excellent means by which to control the final drug purity.

The chromatographer must also contend with the potential products of degradation which accumulate in the drug over time. The presence of light, oxygen, water, heat or various combinations of these factors may cause a chemical reaction of the drug to take place. The degradation products of the drug can possess undesired pharmacological effects for the patient as well as compromise the pharmaceutical elegance of the drug product. The identification of these materials may be determined by one of two methods. One can simply store a sample of the drug and wait for a sufficient amount of time for the products to form. As this can be a very time consuming process, the process may be accelerated by storing the samples under extreme conditions of heat, light, pH, or oxygen. In this manner, the formation of the degradation produces is hastened. They may also be formed in larger amounts than actually occur in a real sample, which allows the chromatographer the luxury of making their detection and separation an easier process. These accelerated samples are subjected to multiple screening processes such as gradient HPLC temperature-programmed GC, or TLC development with a variety of eluant polarities.

Figure 1 is an example of the type of screening procedure which may be applied to a compound of interest. The upper chromatogram is a reversed-phase system for a synthetic intermediate. In this system, the sample appears to be very pure with only a small amount of impurities eluting on the tail of the main peak. The same sample separated in a normal phase system is shown in the lower chromatogram. With the normal pahse system several impurities which eluted under the main peak in the reversed-phase system are completely separated and may be easily quantitated.

Isolation, followed by spectroscopic testing, is then carried out in an effort to identify the degradation products.

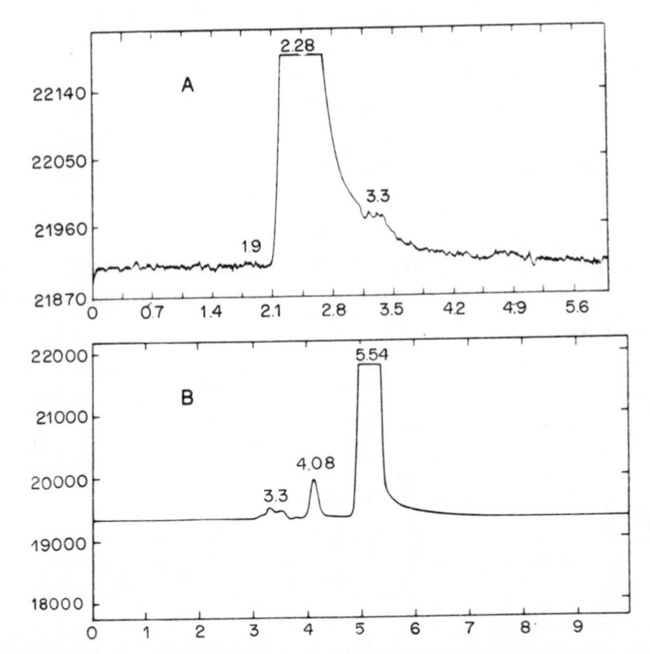

Figure 1. Comparison of reversed-phase (A) and normal phase (B) HPLC chromatograms of a synthetic intermediate.

These products are used to validate the selectivity of the proposed chromatographic assay. Once again, an understanding of the chemical nature of the drug itself is very helpful in predicting the types of degradation products which may form.

In addition to the bulk drug substance, the chromatographer is also called upon to assay for the drug in the drug product itself. This again involves development of a chromatographic method which uses a reference standard for quantitation of the drug. Separation of the potential degradation products is an important consideration as the product will likely be exposed to a variety of environmental conditions during its lifetime. Extreme environmental conditions may promote the degradation of the drug and expose the patient to potential harm. One of the most stringent regulatory requirements faced by the chromatographer in the pharmaceutical industry is the *proof* that his methods are capable of *detecting* and *quantitating all* the degradation products which may form in the product during its shelflife.

Figure 2 shows two HPLC chromatograms of a codeine phosphate and acetaminophen tablet formulation (2).

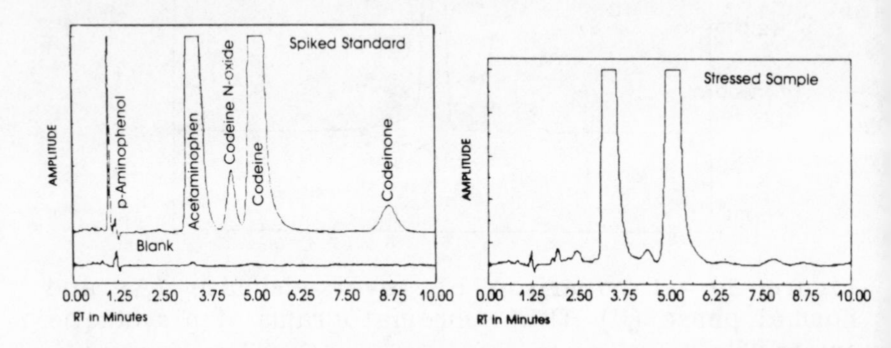

Figure 2. Left: A chromatogram of a standard solution of acetaminophen and codeine phosphate spiked with potential degradation products. Right: A chromatogram of a thermally stressed granulation of acetaminophen and codeine phosphate.

The first chromatogram has been spiked with the known degradation products of the active ingredients. The second chromatogram is of a formulation which has been thermally stressed. Note that several of the degradation products are observed.

The analysis of biological samples presents a very different analytical problem for the chromatographer. In the previous cases, the analyte of interest is usually pure and present in a concentration which does not require significant sample preparation or highly sensitive detectors. Biological samples often contain a small amount of drug and metabolites as well as a significant proportion of uncharacterized and extraneous materials which may interfere with the detection and quantitation of the analytes. The challenge for the chromatographer in this instance is to develop a sample preparation scheme, which will present the sample to the chromatographic system in a form that permits convenient separation of the drug from any remaining extraneous components, as well as any metabolites which may be present. In this manner, an accurate and precise quantitation of the drug may be made. An example of this type of method development (3) is seen in Figures 3 and 4.

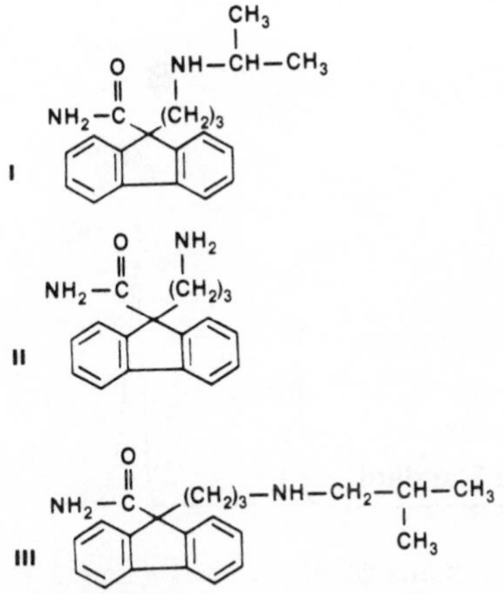

Figure 3. Structures of (I)-indecainide, (II)-its desisopropyl metabolite, and (III)-the internal standard.

The chromatograms in Figure 4 demonstrate the separation of indecainide, its desisopropyl metabolite, and the internal standard. In the first chromatogram UV detection is used. Note that the metabolite elutes close to the peaks from the plasma.

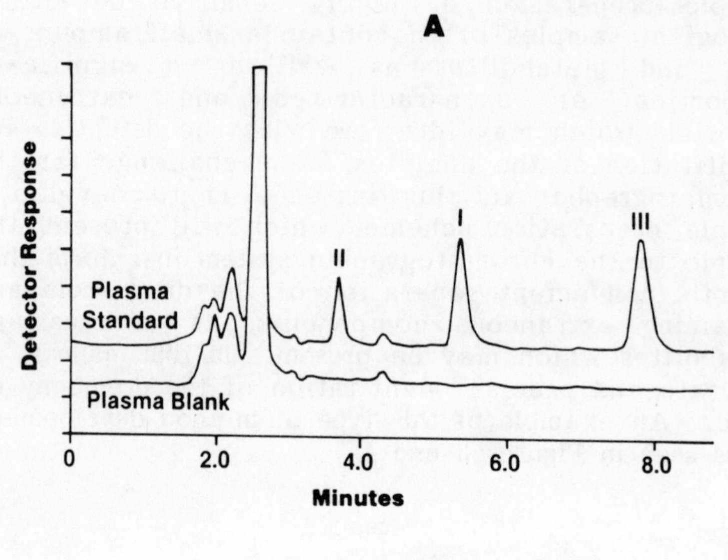

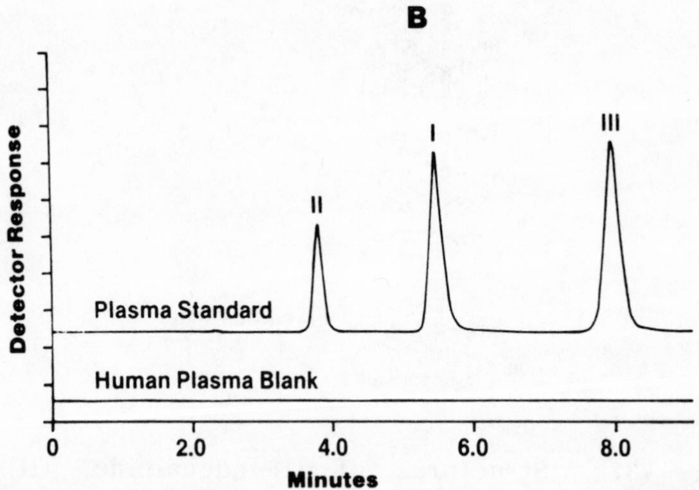

Figure 4. A. A chromatogram of indecainide, metabolite, and internal standard with UV detection. B. The same chromatogram as A but with fluorescence detection. Peak identification: (I)-indecainide, (II)-indecainide metabolite, (III)-the internal standard.

When fluorescence detection is used, however, the plasma peaks are not observed, and an accurate quantitation of all three components may be made. This kind of information, in conjunction with other pharmacological data, is useful in determining the half-life of the drug in the patient and the therapeutic level needed to control the patient's condition. All these factors are important in understanding the nature of the drug's action in the patient and determining the proper dose to be given. Some products have a very narrow therapeutic window where the beneficial properties of the drug outweigh the harmful ones. As the concentration of the drug in the patient increases, the benefits may disappear and a more serious set of problems begin. Clinical laboratories often need to screen patient samples to determine both the identity and levels of unknown or illicit drugs. Chromatographic methods provide a major service in answering these types of problems. There are numerous examples of these methods in the literature (4-6). In fact, there has been a significant increase in the number of published methods for these types of samples as improvements in chromatographic equipment and sample handling procedures have evolved.

A few general statements should be made regarding actual development of the separation conditions. Many times, the trial-and-error method of varying column type, mobile phase composition for HPLC, column temperature for GC, or eluant strength for TLC will lead the analyst to a suitable system. However, as the regulatory requirements that all real and potential components of a drug product be accounted for become a larger part of pharmaceutical method development, the analyst must begin to take a more systematic approach to method development.

There has been much discussion in the literature regarding systematic method development for HPLC.

This approach may have application in method development for bulk drugs. In the instance where the analyst has authentic samples of each of the potential impurities that may be present in the drug, the approach gives the chromatographer the assurance that he is separating all the synthetic precursors and identified degradation products. This set of conditions may also be adapted as the assay method for the individual intermediates if the original method development has been successful in separating all the individual components or the original mixture. If the identity of the degradation products are known as well, the method may also be adapted as the assay method for the drug product. This systematic development provides the assurance that if these degradation products should appear in the product, they will not interfere with the assay of the main component. In this manner, the assay is called a stability-indicating assay.

Often, gradient elution or temperature programming techniques are used to screen a particular sample or synthetic mixture for the components of a given drug. This type of screening can provide valuable information as to the general direction the method development should take by revealing the presence of previously undetected peaks. The analyst should keep in mind the limitation of each method. In both HPLC and GC, there is the possibility that one or more of the components may not elute from the column. Thin-layer chromatography has an advantage in this instance in that the plate will still contain all the sample originally applied after the separation has occurred. For all three techniques, the choice of the proper detection mode is crucial to ensure that all components of the mixture have been detected. In the case of UV detection in HPLC, this problem can partly be overcome by the use of a multiple wavelength scanning detector capable of monitoring several wavelengths simultaneously. This provides some assurance that the UV absorbing components will be detected if present in sufficient quantity. The use of low wavelength detection may also be tried as a method of detecting compounds with a wide range of absorptivities. It will not aid in the case where non-UV absorbing components may be present. Fluorescence or electrochemical detectors may be appropriate in some instances in HPLC as well. Flame ionization detection in GC tends to be a universal detector, but its utility can be limited for detection of minor components. Gas chromatography still commands a significant role in the analysis of pharmaceutical samples. The advent of very high molecular weight products such as polypeptides or thermally-unstable compounds such as the antibiotics limits the scope of the technique, however. New advances in the development of supercritical fluid chromatography as a routine analytical technique hold more promise in solving these problems. The coupling of the supercritical fluid mobile phase and a detector such as the flame ionization or nitrogen/phosphorous detector seems an obvious way to circumvent the lack of universal detection for HPLC and the inability of GC to elute thermally-labile compounds. Thin-layer chromatography can be a powerful tool for screening for unknown materials in drug products and bulk drug substances. By using several developing solvents of widely varying degrees of polarity the chromatographer is able to provide a relatively high degree of assurance that all possible components of the sample have been separated in systems. Once again, the problem of universal detection can limit the scope of the technique. Multiple spray reagents including a "universal" detection technique such as charring may be used to provide a complete detection screen.

SAMPLE HANDLING TECHNIQUES

Pharmaceutical samples are certainly not unique in the problems encountered in sample handling and pretreatment. The proper preparation of the samples can have a large influence on the success of the chromatographic experiment. There are far more details to sample handling than may be covered here. The reader is referred to the general literature. In particular, Snyder and Kirkland (7) have provided a good overview to the subject. The goal of any sample preparation for chromatography is to solubilize the analyte in a homogeneous solution that will not harm the chromatographic system, degrade the sample, or degrade the separation. The types of samples one encounters in this field and the types of sample preparation and handling considerations will now be briefly discussed.

Bulk drug substances and other "pure" compounds such as synthetic precursors generally only require the use of an appropriate solvent for sample preparation. Consideration must be given to the nature of the chromatographic technique so that the solvent choice will not deteriorate the efficiency of the separation. In particular, the efficiency of HPLC separations are known to depend somewhat on the solvent strength of the sample solvent relative to the mobile phase. Use of strong solvents, such as tetrahydrofuran and methanol, in reversed-phase methods where there is a significant aqueous component in the mobile phase will result in chromatograms with poor peak shapes, peak splitting, and loss of resolution. In many instances, the choice of the proper strength sample solvent is simply a matter of selecting one similar to or weaker than that of the mobile phase. Variation of solvent pH or the use of mixed solvents will often be useful in solving problems relating to limited sample solubility. For gas chromatographic separations, sample solvent choice is often dictated by the type of detector to be used or by the injection technique (split or splitless) if capillary columns are to be employed. For particularly impure samples, a filtration step may be required to remove any undissolved material that could plug a column or column-frit.

Sample preparation for *drug products* becomes a little more complicated. The analyte of interest has been formulated with a variety of excipients which must be removed prior to analysis. Simple solubilization of the drug followed by filtration of the insoluble components is again the quickest method of preparation. However, now the analyst must ensure that complete or near-complete recovery of the drug occurs. This can be accomplished by running a set of spiked samples which contain a known amount of drug added to an excipient placebo. Assay of these spiked samples versus a set of standard solutions, provides a measure of the recovery of the sample from the matrix. In the case of complete recovery, future samples may be run versus a set of standards dissolved in the sample solvent. However, when complete recovery cannot be achieved, a correction must be applied to the assay to account for the unrecovered material. A simple approach is to prepare the standards in the sample matrix each time the assay is run. The inherent assumption is that the relative recovery is the *same* for both samples and standards. This technique is useful when the sample matrix varies from assay to assay and unspiked matrix material is available. Any variability in the recovery will be accounted for at the time of each assay. In rare instances, the sample will require that the analyte be solubilized in a solvent other than that used in the actual chromatographic experiment. In these cases, the nature of the sample is such that the preferred injection solvent is inadequate for removal of the drug from the matrix, or the drug is present in very small amounts and must be preconcentrated prior to analysis. This type of problem is becoming more common with the increased development of sustained or extended-release products and much more potent drugs which require low doses to be effective. These sample preparations require that the analyst have a good understanding of the chemical and physical nature of the sample so that a proper choice of action can be taken.

The analysis of *biological samples* is the worst case for sample preparation and handling. Here, the analytes are minor components of the entire complex sample. In contrast to drug products which contain relatively well understood excipients, biological samples contain a variety of unknown proteins, salts, etc. that can seriously hinder the accurate determination of the components of interest. The low concentrations of the analytes may cause an additional problem relating to lack of detection sensitivity. The injection of biological materials, such as dissolved proteins, will have a disasterous effect on the efficiency and lifetime of chromatographic columns. Much work has been done in this area and the reader is referred for example to a recent review (8) in the literature. Here, only some of the common techniques which are used for sample preparation will be covered.

The various techniques which can be used for this type of sample are listed in Table 1.

TABLE 1

SAMPLE PREPARATION TECHNIQUES FOR BIOLOGICAL SAMPLES

1. Protein Precipitation
2. Solvent Extraction
3. Adsorption
4. Concentration
5. Column Switching

The first technique, *protein precipitation*, is most useful for samples which contain sufficient amounts of drug to warrant that no further concentration steps need be taken. Antibiotics are a good example of this type of sample preparation as the doses needed for efficacy provide plasma concentrations in a concentration range sufficient for direct quantitation by HPLC. Simply adding an organic solvent such as acetonitrile and centrifuging to remove the precipitated protein may be sufficient sample preparation. This technique has the advantage that little error is introduced to the assay due to the small amount of sample handling that occurs. A final filtration step should also be added to ensure that no particulates are injected into the chromatographic system.

Solvent extraction involves extracting the analytes from the original sample into an organic phase. As the biological samples will consist of primarily water, this step provides an opportunity to exclude the hydrophilic components of the sample and clean up the sample prior to analysis. Varying the pH of the aqueous phase can cause the selective extraction of ionized or unionized compounds. A second extraction back to an aqueous phase may be needed to remove further unwanted materials. Variables which are often tried include the pH and ionic strength of the aqueous phase, the polarity of the organic phase, and the volume of each phase. This technique can introduce significant errors into the quantitation if there are sample losses or incomplete extraction.

Adsorption is similar to the solvent extraction technique. The difference, here, is that one portion of the sample is "extracted" or adsorbed onto a solid surface. This technique has enjoyed a revolution in recent years with the commercial availability of prepacked adsorption materials such as the Sep-Pak™ cartridges (Waters Associates) and the Bond-Elut™ cartridges (Analytichem International). These cartridges contain a small amount of silica or bonded silica in a disposable cartridge. In the case of the reversed-phase packing, the sample is introduced with a very polar solvent such as water. The polar components are eluted while the non-polar components are retained on the cartridge. The sample components may then be eluted by changing to a slightly less polar solvent, while leaving the more strongly retained materials on the column. These cartridges are now available in a variety of packing materials including reversed-phase, amino, diol, and underivatized silica. The convenience of these cartridges has made preparation for this type of sample much easier. There are no concerns about cross contamination of samples because they are used only once and they provide a very clean, efficient separation once the correct set of solvent washes have been found. Another version of this method uses a disposable precolumn in front of the analytical HPLC column. This technique is effective only if the undesired components remain on the guard column and do not slowly elute onto the analytical column. The guard column is then changed once its capacity has been exhausted. There is an inherent danger with this particular technique in that unknown or unwanted materials may carry onto the analytical column and plug the inlet frit or degrade the column efficiency.

Concentration of samples may be carried out in conjunction with both solvent-solvent extraction and adsorption techniques. The analyst may use a large quantity of sample and after preparation by either extraction or adsorption, simply remove the sample solvent and reconstitute the sample in a small quantity of solvent to be injected. The adsorption cartridges are particularly effective for this type of technique as they may be easily loaded with a large quantity of sample prior to elution of the sample components. The concentration step also allows the analyst to dissolve the analyte in a solvent appropriate for the chromatographic experiment without being restricted to the use of that solvent for the sample cleanup procedure. This concentration step in combination with the other sample preparation techniques has allowed chromatographers to quantitate drugs which are present in the original sample in very low concentrations. The sample extraction/concentration steps may also be automated so that the analyst is not limited by the number of samples that can be manually prepared. The advent of laboratory robots with the capability of handling relatively complex sample preparation (see Chapter 8) holds much promise for the clinical laboratory.

Column switching techniques accomplish the same type of sample cleanup as the adsorption method cited above, but it is done as part of the chromatographic experiment. Here, the sample is introduced to a precolumn via an injection loop. The sample on interest is then eluted from this precolumn onto the analytical column. A switching valve between the precolumn and analytical column is used to control the elution of the analytes onto the analytical column as well as removal of the undesired components to waste. The injected sample then continues through the analytical column. Another variation of this method involves samples with a range of polarities. The sample adsorbed on the precolumn is rinsed with a mobile phase of low strength, and the less retained components are eluted onto the analytical column. Once they have been separated, a switching valve between the precolumn and analytical column is turned so that the eluant from the precolumn is sent to a second analytical column. A stronger mobile phase is then used to elute the adsorbed components onto the second analytical column for separation and quantitation.

The proper use of sample handling techniques is as an important part of the development of a good quantitative method as the choice of the proper chromatographic system. Good laboratory technique, and an understanding of the influence of sample preparation on the precision and accuracy of a given method, will play an important role in developing a rugged method which does not require continued redevelopment.

FACTORS DETERMINING THE ACCURACY AND PRECISION OF QUANTITATIVE ANALYSIS

The utility of any chromatographic method is limited by both its ability to separate and detect the species of interest and its ability to perform this task in an accurate and reproducible manner. A method which generates reliable results only part of the time is certainly of little use. A variety of factors come into play when using a chromatographic method to produce quantitative data. These factors include the ability of the chromatographic system to perform the desired separation, as well as the nature of the samples and their method of preparation prior to the analysis. Technological improvements in the quality of instruments and detectors, and in the manufacture of stable, durable, and reasonably reproducible columns, have enabled chromatographers to expand the range of compounds that may be conveniently assayed. Sample preparation techniques have likewise become more sophisticated to the point that some of them actually perform a selective pre-chromatography "chromatographic" separation. Many of the details concerning the actual performance of columns, detectors, and sample introduction techniques have been discussed elsewhere. This discussion will be focused on the factors to be considered when developing and validating a method for quantitation by a chromatographic technique.

Validation of chromatographic methods is a topic which is not limited to pharmaceuticals. The validation tests which are discussed here are applicable to all sample types and all chromatographic techniques. Much validation work is done with these methods in the pharmaceutical industry due to requirements initiated by various agencies which monitor the development, manufacture, and sale of pharmaceutical products (9-11). These requirements are part of the larger concern that only safe and effective products are made available for sale to the public. Ensuring that the analytical methods for assay and control of these products are operating correctly is one way to protect the safety of the patients using the product.

The parameters which are most commonly evaluated in the development of a method are listed in Table 2.

TABLE 2

ASSAY VALIDATION TESTS FOR ANALYTICAL METHODS

1. Linearity
2. Precision (Ruggedness)
3. Recovery (Accuracy)
4. Specificity
5. Detection Limits

Although this list is not all-inclusive, it covers the major points that should be considered. Much of the literature discussion in this area has focused on the validation of HPLC methods due to the explosive growth of this technique. These discussions have included the topics of actual validation procedures to be used and the criteria to be applied to decide if a method is acceptable for use. In most cases the validation principles apply to GC and TLC methods as well. An understanding of the nature of the sample, its matrix and the relative concentration of the analyte should be taken into account when applying acceptance criteria for a validation test. It is unreasonable to expect a method for quantitation of a drug in ppm quantities in a complicated matrix to be performed with the same precision and accuracy as the simple assay of a pure drug substance. Choice of the acceptance criteria is generally left to the analyst, although methods for drug substances and drug products have had criteria set for them by the compedia (12). These items will be briefly discussed as they pertain to pharmaceutical assays.

Linearity

The *linearity* of the method is tested to ensure that signal response for a sample is directly related to its concentration. A linear response enables the analyst to use linear regression techniques to calculate a standard curve which is then used to calculate concentration data for the samples. Non-linear standard curves can be used, but they are limited by the accuracy of the regression analysis used to calculate the standard curve. The range of the standard curve may be narrow (50-150% of nominal concentration) for methods where the sample response is easily anticipated, such as drug substance and drug product assays. Alternatively, a range of two orders of magnitude or more may be needed for biological or environmental samples where the analyte concentration may vary by a wide margin. Acceptance criteria can include a minimum value for the coefficient of determination, or a limit for the value of the maximum percent deviation from the calculated line for any given point.

Precision

Method *precision* is a measure of the reproducibility of the method. This is generally determined by assaying multiple replicates from a homogeneous sample and calculating the standard deviation or relative standard deviation of the results. The precision of the method is a combination of several factors including the homogeneity of the samples, the sample preparation techniques, and the actual separation and detection of the components. The use of real samples when available enhances the validity of this determination. One should note that methods which require significant sample handling and preparation prior to analysis will probably exhibit worse precision than a "dilute-and-shoot" method. There may be occasions, however, where the extra sample cleanup will lead to improved method precision due to significant improvements in the separation and detection of the analytes. Methods which will ultimately be transferred from one laboratory to another may undergo a multiple-lab, multiple-analyst, multiple-day, multiple-column evaluation to determine the actual variability of the method over time. This determination is often called the method ruggedness. It is used to estimate the actual variability of the method as it is used over a period of time. The ruggedness of a method is usually determined at a later time that the initial method validation. It should be noted however, that the determination of the column-to-column variability should be determined at the time of the initial method validation. Differences in columns from lot to lot can have significant effects on the reproducibility of the method and can lead to having to totally redevelop the chromatographic conditions (13).

Recovery

Recovery is simply a determination that all the species of interest are carried through the sample preparation scheme without loss. This determination is also occasionally referred to as the accuracy of the method. Recovery is determined by assaying samples of matrix to which known amounts of the analyte have been added and comparing results of their assay to those of standards prepared without matrix. The recovery is the amount recovered from the spiked samples compared to that from the standards. Once again, the acceptance criteria for recovery must depend on the nature of the samples and the concentration of the analytes. Low recoveries can be accounted for by simply preparing the standards in a blank matrix and carrying them through the identical sample preparation as the samples. One must assume, however, that this simple spiking experiment sufficiently mimics the environment of the analytes in the samples.

Specificity

Specificity refers to the ability of the method to resolve the species of interest from all other components of the mixture. This is determined by spiking a solution with all the known possible interfering components of the sample. In the case of drug substances and drug products, these would include the synthetic precursors and degradation products. For biological samples, it may be necessary to test the method for potential interferences from other medications taken by the patient. Coelution of any of these interfering species will certainly have an effect on the precision and accuracy of any chromatographic method.

Detection Limit

The determination of the *detection limit* for an analytical method is not a requirement for all methods. It is carried out when the species of interest may be present at very low concentrations, as in the case of the determination of a degradation product of a drug. This measurement may be done with varying degrees of rigor depending on the method and the identity of the analyte. Often, a simple determination of the instrument detection limit is sufficient to identify the smallest amount of material that may be conveniently detected. In the case of a compound with important pharmacological or regulatory considerations, a more extensive determination of the detection limit may be done to provide a better estimation of the day-to-day variation in the detectability of the species of interest.

Stability

Other tests which merit mention include determination of the *stability* of the sample solutions and matrix effects. The extended use of auto-injectors for HPLC and GC assays has certainly increased the number of samples that one analyst may run in a given length of time. However, these automated systems are not infallible, and occasionally a problem will occur while the instrument is running unattended that will prevent the completion of a run. In this event, it is useful to know if the samples and standards may simply be revialed and run again. Testing the stability of these sample solutions provides an assurance that the integrity of the analytes has not changed. This test may be run at the time of determination of the method precision by simply running the samples again after an 18-24 hour interval and comparing the response for the aged samples to that of the fresh samples. This length of time provides a comfortable margin of safety in the event of an interruption in the assay.

Matrix Effects

Matrix effects may be evaluated when the analyte of interest is present in samples at low levels. This is tested by calculating standard curves for a set of standards and a set of standards spiked into the sample matrix and comparing the resulting slopes. If the slope of the spiked standards significantly differs from that of the standards, one may then use standard addition methods or spiked standards to account for this difference in response for the samples.

System Suitability

The use of *system-suitability* tests for chromatographic assays has been recommended by both the FDA and the compedia (12). Unlike the method validation that is carried out only once at the development stage of a method to show that the method is suitable for use, a system-suitability test is run at the time of each use of the method. The purpose of this test is to demonstrate that the chromatographic system is properly operating at that particular time. This generally entails a test of both the precision and chromatographic performance of the system. The precision is demonstrated by the injection of one standard or sample solution five or six times. The response of the peak of interest is measured, and the reproducibility of these responses is calculated. Generally, a limit of not more than 2% RSD is considered acceptable performance. The second part of the system-suitability test includes the determination of a chromatographic parameter such as the resolution factor for two peaks or the tailing factor for an important peak. The resolution should be calculated between the two peaks most difficult to separate in the chromatogram. For methods where no extraneous peaks are expected, the resolution may be calculated between the analyte peak and the internal standard peak. Other chromatographic parameters may be used as system-suitability criteria, but it is important to remember to choose a test that is a relevant indicator of the performance of the system.

EXAMPLES

Numerous examples of the principles discussed here may be found in the literature. A few recent papers will be examined to demonstrate the use of chromatographic methods in the pharmaceutical field and the factors which were important in developing good quantitative methods.

The assay for the *active ingredient* in a drug product is often a relatively simple task for the chromatographer. In the case of a proprietary drug sold by one manufacturer, the method development usually involves only one formulation and one set of excipients. Development of the assay for the active ingredients for generic or over the counter medications is a more difficult problem, as each manufacturer may use a different set of excipients or add additional ingredients to the product. One example of this type of problem was recently discussed (14). The authors' goal was to develop an assay for dextromethorphan [(+)-3-methoxy-17-methyl-9a,13a,14a,-morphian hydrobromide monohydrate], an active ingredient in many commercial cough and cold syrup medications. The method needed to be precise, rapid, and sensitive, as well as applicable to the variety of formulations which currently contain this ingredient. Both HPLC and GC were tried. Adequate chromatographic conditions were developed for each technique (Figures 5 and 6).

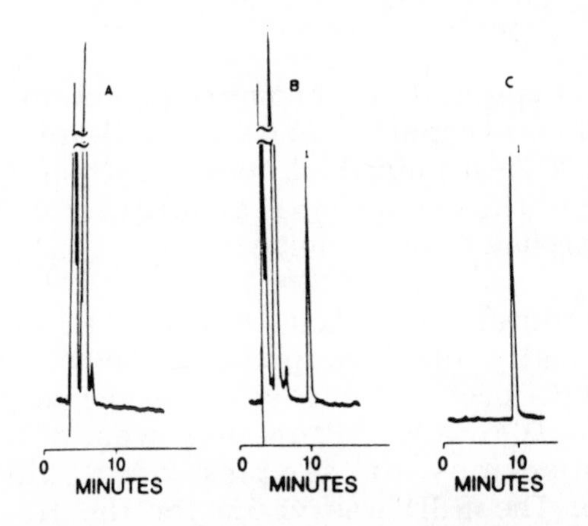

Figure 5. Typical HPLC chromatograms for (A)-placebo syrup, (B)-sample syrup, and (C)-standard. Peak 1 is dextromethorphan hydrobromide.

The HPLC sample preparation involved simply diluting the sample in the mobile phase prior to chromatography. For the GC assay, the samples had to be extracted first. Dextromethorphan is used as the hydrobromide salt and must be converted to the free base prior to GC analysis. A simple extraction scheme was developed where the sample was added to a 0.1 M NaOH solution and extracted with three portions of dichloromethane. The internal standard was added, and the sample was diluted to a specific volume prior to analysis.

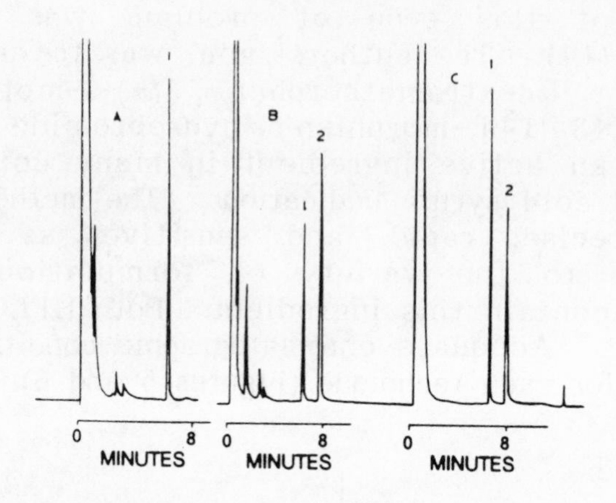

Figure 6. Typical GC chromatograms for (A)-placebo syrup with internal standard, (B)-spiked placebo syrup, and (C)-standard. Peak 1: O-methoxyphenyl benzoate (internal standard), Peak 2: dextromethorphan hydrobromide.

Recovery studies indicated that the HPLC procedure gave essentially ideal recoveries (100.1 and 100.2%), while the GC procedure recoveries were not as good (103.8 and 104.5%). Precision was determined by multiple injections of samples of a single lot of cough syrup. The RSD was <1.5% for the HPLC procedure and <4.0% for the GC assay. Further studies involving the assay of commerically available cough syrups indicated that the HPLC assay yielded data very close to the label amounts on the product while the GC results were consistently high. The high values were attributed to unusual partitioning behavior of the dextromethorphan in the organic layer rather than to sample matrix effects. The conclusion of the authors was that the HPLC assay was the preferred method due to its simple sample preparation, better recovery and precision, and shorter overall assay time.

The assay of *insulins* has traditionally been carried out using biological assays (15). These assays were developed prior to the advances in modern chromatographic methods, and are therefore logical candidates for method revision. The authors of a recent publication (16) had several goals in mind for this assay. They desired a method which would be a universal method for all types of insulin, both human and pancreatic. They wanted to achieve stable sample solutions that could be used in conjunction with an auto-injector to allow for unattended assay. Also, in the method designed all sample solutions were of the same concentration such that one set of standards could be used for any type of sample. Sample preparations schemes were tested to find one set of conditions that would allow for maximum sample stability after preparation. A mobile phase with low pH was known to be required to achieve adequate peak shape in the HPLC assay, but insulin solutions at low pH have limited stability. Dilution in mobile phase required that the samples be acidified just prior to analysis. This made the use of auto-injectors for overnight operation nearly impossible. Experiments showed that a pH of 9 in conjunction with the addition of EDTA to the sample solutions provided very good sample stability to the point that standards could be prepared, refrigerated, and used for several months with no measurable loss in potency. The utility of the proposed sample preparation scheme was tested for the various insulin formulations as well as the bulk insulin crystals and was found to be satisfactory. A method validation was carried out which tested the linearity and accuracy of the method. The ruggedness was estimated with an interlaboratory study in which identical insulin samples were assayed by analysts in several laboratories. An example of the separation of the different insulins is shown in Figure 7.

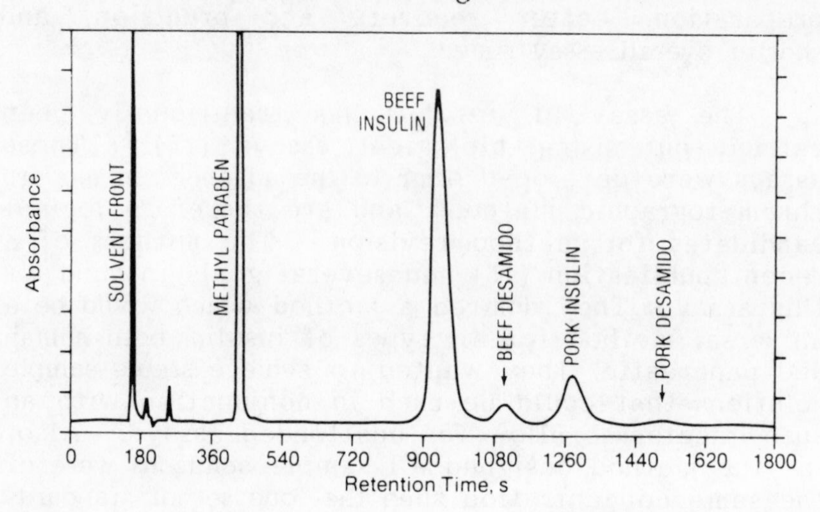

Figure 7. A chromatogram of a representative sample of a mixed beef/pork Lente insulin formulation.

Table 3 lists some of the precision data generated in this study.

TABLE 3

Estimated Relative Standard Deviation for HPLC Assay Results Within a Laboratory

Formulation	Number of Replicates		
	1	2	3
NRI	1.9%	1.8%	1.7%
NPH	2.3%	2.0%	1.9%
Lente	1.8%	1.7%	1.7%

One conclusion which may be drawn from these data is that the multiple replicates will ensure a more precise result for any of the three formulations. However,it is worth noting that there is not a significant improvement in any of the three cases by increasing the number of replicates. This indicates that in the interest of saving time, the number of replicates for a sample may be reduced without a significant loss of precision of the data.

One other consideration for the authors was how this method would compare to the current compedial assay for insulin (12). The results for 21 lots of human zinc crystals are outlined in Table 4.

TABLE 4

Comparison of HPLC Potency and Rabbit Bioassay Results for Human Insulin Zinc Crystals

	Average Potency	RSD	Range
Biopotency (U/mg)	28.06	2.8%	26.4-29.1
HPLC Potency (U/mg)	28.33	1.0%	27.8-29.0

These data indicate that the results generated by the HPLC method are certainly within experimental error of those from the bioassay and that the HPLC method is more precise as well. The conclusions of the authors were that HPLC will ultimately replace the use of bioassays for insulin and that these chromatographic assays will provide a faster and more precise determination.

While there is certainly interest in publishing chromatographic methods for bulk drug products, it does not seem to compare to the interest in methods for the assay of *drugs in biological matrices.* These methods provide a valuable source of information for other workers in the field as new products are tested in both animal models and human subjects. Improved analytical methods for drugs which are already on the market may find utility in clinical laboratories faced with assaying large numbers of samples on routine basis.

As mentioned earlier, the goal for assays of biological fluids is to determine the levels of both the parent drug as well as the metabolites of the drug. The authors of the recent work (17) described a method for the determination of the new antihypertensive drug, pinacidil, and its major metabolite in plasma. The authors described two sample preparation schemes and two slightly different chromatographic systems which were used.

The first method was designed for the assay of the parent drug only. The sample preparation in this case was a simple protein precipitation by addition of acetonitrile followed by centrifugation, evaporation, and reconstitution. Later experiments determined that this preparation would not allow an accurate quantitation of the N-oxide metabolite due to low extraction efficiency from aqueous media. The second scheme involved cleanup of the sample on a Sep-Pak™ C-18 cartridge. After the undesired components of the sample were eluted from the cartridge with a water rinse followed by an 80:20 water:methanol rinse, the analytes were eluted with methanol, taken to dryness, and reconstituted in mobile phase.

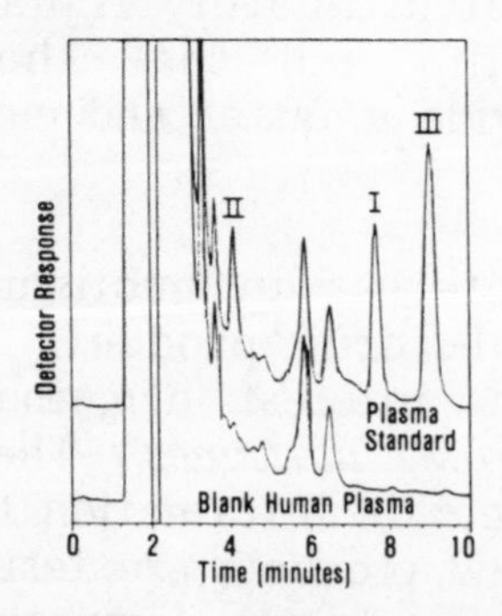

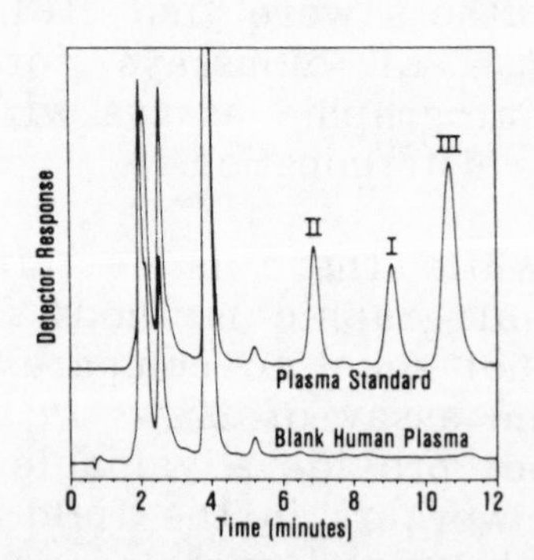

Figure 8. Typical HPLC chromatograms of plasma samples obtained by Method A (left) and Method B (right). Peak identification: (I)-pinacidil, (II)-pinacidil N-oxide, (III)-the internal standard.

Representative chromatograms of standard solutions prepared in this manner are shown in Figure 8. Note that the peak for the N-oxide in the first preparation is not well resolved from the plasma components which elute early in the chromatogram. Further work with these compounds indicated that in the mobile phase buffer pH (7.0) used for the first chromatogram, the N-oxide has very poor absorbance in the UV at 254 nm, the wavelength of determination. By changing the pH to 4.0, the UV spectrum of pinacidil red shifts such that both compounds have similar absorption at 284 nm. This enabled their simultaneous determination in one chromatogram. The new mobile phase was used for the second sample preparation and the results are also demonstrated in Figure 8. In this case the N-oxide was well resolved from the early eluting components and could be easily measured. Precision studies for both these methods indicated that the second scheme provided slightly better precision in spite of the increased sample handling. This was attributed to better sensitivity for both compounds at the low pH and different wavelength, as well as to improved chromatography which permitted better peak quantitation. The overall limit of detection for pinacidil was estimated to be 10 ng/mL for the first method and 5 ng/mL for the second. Recoveries were measured by spiking experiments and were determined to be 100% for the first preparation and greater than 90% for the second. The authors concluded that the first method was viable, particularly in the instance where large numbers of samples needed to be assayed for the parent drug only. By taking advantage of the pH dependence on the UV absorption curve for pinacidil, they were able to find chromatographic conditions which allowed simultaneous determination of both the parent drug and its major metabolite in one chromatographic experiment.

Not all biological samples are as easily

assayed, however. A recent paper (18) described the development of the assay for a new oral antidiabetic drug and its four hydroxylated metabolites. Due to the large polarity differences in the analyte, it was not possible to develop a single isocratic HPLC system that would resolve all five components. The problem was solved with a column switching assay which used two isocratic systems with different polarity mobile phases. A schematic diagram of the system is shown in Figure 9. After cleanup of the samples with both reversed phase and silica Bond-Elut™ columns, the reconstituted samples were injected onto the 3-cm guard column packed with a C-18 modified silica.

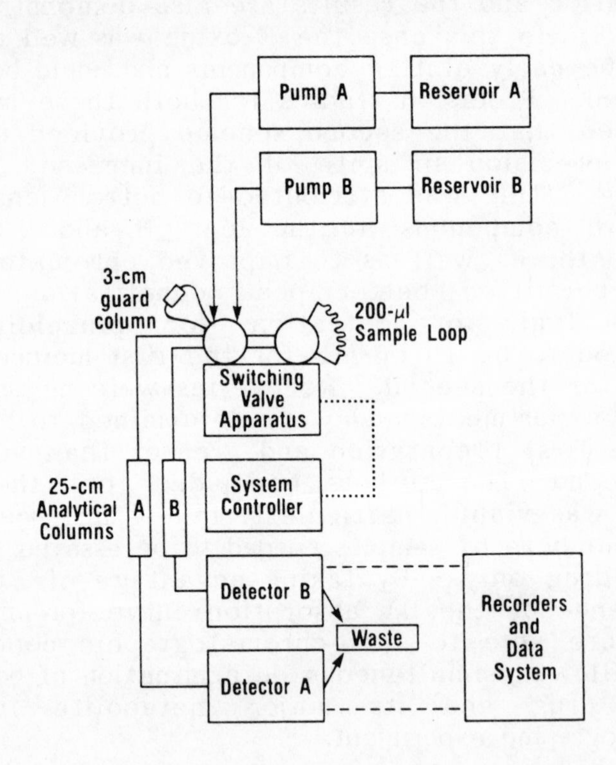

Figure 9. Schematic representation of the HPLC apparatus for isomodal column swtiching.

The initial mobile phase which was passed through the precolumn was 40:60 acetonitrile/phosphate buffer pH 2.5. Under these conditions, the hydroxylated metabolites were eluted onto the analytical column for system A, and the parent drug and its internal standard were retained on the precolumn. After the metabolites were on the analytical column, the switching valve was turned, and a mobile phase consisting of 70:30 acetonitrile/phosphate buffer pH 2.5 was pumped through the precolumn to elute the parent drug and the internal standard onto the system B analytical column. An example of the chromatograms resulting from this system is shown in Figure 10.

The success of this chromatographic system was due to the large differences in k' between the metabolites and the parent drug. By using two isocratic HPLC systems and only one sample preparation and injection, the system was found to be very reproducible and rugged. A time saving was also realized with this switching system as only 25 minutes were required to assay for one sample versus 45 minutes for a gradient method capable of resolving all the sample components.

A final example, taken from the literature, discusses the determination of the R- and S-enantiomers of tocainide, an orally active antiarrhythmic drug which is administered as the racemate (19). Other methods for the assay of tocainide in plasma samples had been described previously. The goal of the authors was to develop sample handling conditions which required the use of a smaller sample in addition to lowering the detection limit. Conversion of the tocainide base to the heptafluorobutyryl derivative allowed detection with an electron capture detector after separation by capillary GC.

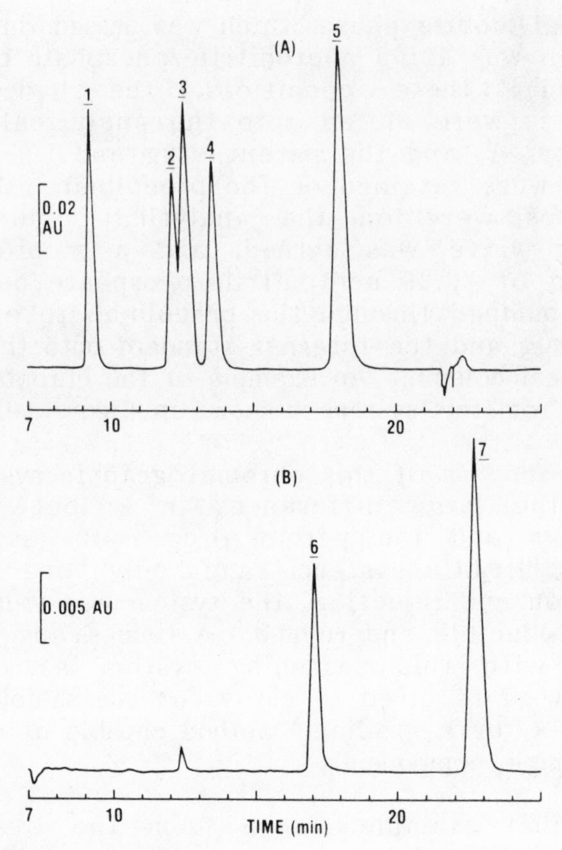

Figure 10. Simultaneous chromatograms collected during the analysis of a solution standard. Chromatogram A, peaks 1-4 are metabolites and peak 5 is the internal standard. Chromatogram B, peak 6 is the parent drug, and peak 7 is the internal standard.

The authors optimized the derivatization conditions by varying the concentration of heptafluorobutyryl anhydride and the nature of the reaction solvent. They found that low concentrations (0.03%) of the anhydride were needed to prevent degradation of the derivative after reaction. It was also easier to remove unreacted anhydride by evaporation prior to chromatography and thus not overload the detector.

The heptafluorobutyryl derivative was easily separated and detected on a methyl silicone capillary column down to levels corresponding to 58 ng/mL of plasma. Injection of the derivatized sample onto a Chirasil-Val capillary column resolved the enantiomers of both the parent drug and the internal standard (See Figure 11). This permitted the determination of the individual enantiomers of the parent drug in plasma samples. Further studies indicated that the heptafluorobutyryl derivative was superior to other fluoro derivatizing agents as determined by the separation of the individual enantiomers. Thus, the careful choice of derivatization reagent and reaction conditions led to an improved assay procedure as well as the ability to resolve and quantitate the individual enantiomers.

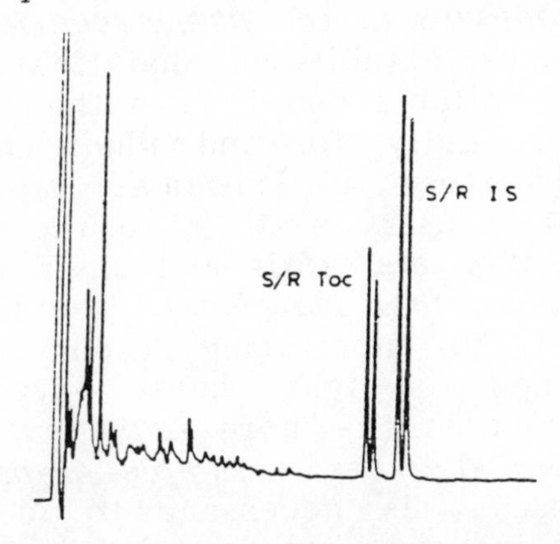

Figure 11. A GC chromatogram demonstrating the separation of the enantiomers of tocainide (Toc.), and the internal standard (I.S.).

Chromatographic techniques have become a significant tool in the analysis of pharmaceutical products. The ability to accurately and precisely quantitate drugs, their impurities and degradation products has led to the manufacture of purer, safer products for the consumer. Analysis of biological fluids for drug and metabolite content in conjunction with physiological data enables the determination of the mode of action and efficacious dose range for new products. Chromatographic methods have been replacing older, less specific or less accurate methods for some time, and it is likely that this trend will continue. It is the *responsibility of* the chromatographer to understand the limits of this powerful tool and perform the best possible method validation. This will ensure that the most accurate and precise methods will be used to assay and control these important products.

SYNOPSIS

The *selectivity, sensitivity* and *convenience* make *chromatography* one of the *most valuable analytical techniques in the pharmaceutical industry.* In most instances, established analytical procedures are not often available and thus methods have to be developed specifically for individual samples and products. The use of chromatography in the pharmaceutical industry extends to the *analysis of biologically active materials* and their metabolites *contained in biological matrices.* The technique is used extensively for monitoring product degradation in the presence of light, heat, oxygen and at different pH. One of the more *important aspects of chromatographic analyses is sample preparation.* In many instances, it is necessary to identify both traces of the intermediate used in the synthesis and the degradation products *as well as the drug itself.* Consequently, good resolution is essential. In the quantitative analysis of pharmaceutical and biological matrices, *selective and specific detectors are popular* to *reduce the demand for high resolution.* In the development of methods, it is extremely important to establish the precision, the accuracy, the repeatability and the "ruggedness" of the overall procedure. Any method must be validated between samples, between laboratories, between analysts, between companies and over extended periods of time. One of the most *important aspects* of a chromatographic analysis of components *in difficult matrices is the recovery efficiency.* The accuracy depends on high level recovery which must be constant and repeatable. Some examples of the use of chromatographic analytical techniques in the pharmaceutical field include the analysis of cough syrup containing dextromethorphan hydrobromide, the assay of insulins, both human and pancreatic, and the assay of the antihypertensive, antidiabetic and antiarrhythmic drugs and their metabolites in biological matrices.

References

1. J. Kirschbaum, S. Perlman, J.Joseph, and J. Adamovics, J. Chromatogr. Sci., 22 (1984) 27.

2. W.R. Sisco, C.T. Rittenhouse and L.A. Everhart, J. Chromatogr., 348 (1985) 253.

3. K.Z. Farid, A.F. Fasola and J.F. Nash, J. Chromatogr., 337 (1985) 329.

4. "Clinical Liquid Chromatography", P.M. Kabra and L.J. Marton (Eds.), CRC Press, Cleveland, OH, 1984.

5. "Methodology for Analytical Toxicology", I. Sunshine (Ed.), CRC Press, Cleveland, OH, 1975.

6. "Analysis of Drugs in Biological Fluids", J. Chamberlain (Ed.), CRC Press, Cleveland, OH.

7. L.R. Snyder and J.J. Kirkland, "Introduction to Modern Liquid Chromatography", Wiley, New York, 2nd ed., 1979 .

8. A.C. Mehta, Talanta, 33 (1986) 67.

9. A.J. Vanderwielen and E.A. Hardwidge, Pharm. Tech., 6(9) (1982) 66.

10. S.M. Sicarro and K.A. Shah, Pharm. Manuf., 3(4) (1984) 25.

11. J.E. Haky and E.A. Domonkos, J. Chromatogr. Sci., 23 (1985) 364.

12. USP XXI The United States Pharmacopia/The National Formulary NF XVI, U.S. Pharmacopeial Convention, Rockville, MD, 1985.

13. I. Wouters, S. Hendrickx, E. Roets, J. Hoogmartens and H. Vanderhaeghe, J. Chromatogr., 291 (1984) 59.

14. V. Gibbs and Z. Zaidi, J. Pharm. Sci., 73 (1984) 1248.

15. G.A. Stewart, Analyst, 99 (1974) 913.

16. H.W. Smith, Jr., L.M. Atkins, D.A. Binkley, W.G. Richardson and D.J. Miner, J. Liq. Chromatogr., 8 (1985) 419.

17. M. Hamilton, K.Z. Farid and D.P. Henry, J. Chromatogr., 375 (1986) 359.

18. J.W. Cox and R.H. Pullen, J. Chromatogr., 307 (1984) 155.

19. A.M. Antonsson, O. Gyllenhaal, K. Kylberg-Hanssen, L. Johansson and J. Vessman, J. Chromatogr., 308 (1984) 181.

Quantitative Analysis using
Chromatographic Techniques
Edited by Elena Katz

Chapter 8

IS AUTOMATION THE FUTURE OF QUANTITATIVE CHROMATOGRAPHY?

G. Raymond Miller

THE NEED FOR AUTOMATED ANALYSIS

Most chromatographers would probably agree that automation is highly desirable. Depending upon the circumstances, automation is expected to provide reduction in costs and errors, greater sample throughput, more reliability, and improved job satisfaction.

While automation does hold many promises, greater benefit may also be achieved through optimization of the analytical method. *Automation* should only be considered whenever the method is performing at a level where *further* attempts at optimization would not be practical.

There are probably very few labs which would not benefit from automating their chromatographic analyses in one form or another. The advantages of automation can be broken down into four categories: *time*, *reproducibility*, *reliability*, and *cost*. The disincentives for automation relate mostly to initial capital and development costs and the pay back period in comparison to the lifetime of the equipment or of the project.

The *time* spent by chemists attending to chromatographic analyses is a prime target for automation. In quality control environments, the analyst's efforts are mainly concentrated in the sample preparation and data analysis phases of the procedure. Any changes in the system which leave the analysis intact but reduce the attention required by the chemist provide time for other activities. If that time is used effectively, then the result is an increase in the productivity of the laboratory.

Sample throughput is another target of automation. The time necessary for an optimized chromatographic separation will be a constant. The only way to increase the number of analyses performed in a given period of time is to begin each run immediately after the previous run has been completed and to run for as long as possible. An example of increased throughput would be in the use of autosamplers and data systems which process the data while the next sample is separating. Each sample is injected as soon as the data system indicates that the previous sample's peaks have been detected.

A laboratory which operates only one shift per day can see dramatic increases in throughput by loading the autosamplers during the day and allowing separations to run during the day and into the night. The use of automatic sample preparation systems should also increase throughput since preparation can occur continually through the night, if necessary.

The total analysis time, time that an analysis takes from receipt of sample to the report of the final result, may also be reduced via automation. This differs from throughput time savings, because throughput can be improved most by running tasks concurrently; preparing one sample while separating another, etc. Total analysis time can only be reduced by shortening the time required for a single step in the analysis. Assuming that the method has been optimized, the two areas where automation will be most effective will be sample preparation and data analysis. In most cases, however, the sample preparation step cannot be shortened, since current sample preparation systems are performing chemical and physical operations and are not necessarily faster than humans. The data analysis stage is very amenable to automation. Faster microcomputers and storage systems and improved software along with online specification systems allow chromatograms to be evaluated rapidly, compared with specifications, and the results to be stored and reported. This can provide tremendous savings in time and effort compared with manual interpretation of the data from strip-chart recordings.

Another advantage of automation is *reproducibility*. Properly operating equipment will give consistent performance; the function will be performed the same way every time. This is particularly important in the areas of sample preparation and sample introduction.

Sample preparation systems generally transfer the sample from one container into another more suitable for introduction into the chromatographic system. They also usually include some physical or chemical actions such as weighing, mixing, diluting, heating, cooling, filtering, or dissolving. The use of an automated sample preparation system helps to insure that all samples processed are mixed the same way and for the same length of time, diluted the same, filtered the same, heated the same, etc. Since the same sequence of events occurs for all samples, the timing of additions, mixing, and the steps necessary to complete the sequence is more likely to be similar between samples than under human control.

Similarly, in sample introduction, the reproducibility of the introduction step helps to maintain reproducibility in the peak shape. For GC autosamplers, the length of time that the needle is in the injection port, the speed at which the sample is delivered, and the depth that the needle is inserted, all may have an effect on the reproducibility of the component peaks. The placement of sample on a TLC plate also has an effect on the final appearance of the developed spots (1).

The injection step in LC is less critical since the injection valve handles the introduction of the sample into the mobile phase.

Reliability is also an advantage of automation. Properly designed and validated hardware and software can insure that each step in the method takes place correctly. Automated systems are not prone to the type of mistakes that humans make: they don't get 'confused', they don't 'daydream', they don't 'get in a hurry' or take 'short-cuts'. On the other hand, if they are instructed in error, they will consistently perform according to the instructions; *in error*. Automated systems are generally ideal for validation of the methodology in use. Multiple samples can be analyzed with assurance that they were run under the same set of conditions.

Finally, *cost* is one of the most important advantages of automated chromatography systems. It is usually the numerical justification for implementing systems providing the advantages described above. If the implementation of automation equipment results in quicker answers, more throughput, better reproducibility, or more reliability, and the result is savings of money in a short enough period of time to make the investment pay off, then the project is worthwhile.

There are many types of laboratories which use chromatographic techniques. In general, production control labs need results from sample analyses in short periods of time. The needs of the laboratory determine the type of automation which may be most beneficial.

A production laboratory where the chromatographic process is the product, such as the separation of monoclonal antibodies from contaminants, might need the capabilities of automatic sampling, gradient control, and fraction collection (2). A forensic lab might benefit from a data analysis system which incorporates an index of retention times and column data for identification of compounds and which might recommend optimum conditions for the separation of the mixture. A quality control lab might be aided by a system which could store a variety of methods for instrument setup and would analyze the resulting data, compare the result with specifications and report the results to an external computer system. An environmental monitoring lab might benefit from a sample preparation system which could prepare wastewater samples for chromatographic analysis, or a method development lab might use a system which automatically repeats separations under different mobile phase conditions until an optimal separation is obtained.

There might well be as many different requirements for automation as there are laboratories using chromatography. A suitable approach to determining the applicability of automation in a variety of situations would be to identify the areas where automation is available and then determine if one of those options would be appropriate.

Sample Preparation

Automatic sample preparation systems are becoming more widespread. Before the sample can be introduced into the separation system, it must be prepared suitably. Only rarely can a sample be injected or spotted without preparation. The preparation might entail dilution, filtration, heating, cooling, derivatizing, mixing, or a variety of other manipulations. The reason for modifying the sample might be to prevent overloading the column, concentrating the sample so that it may be detected, to remove interfering components, to remove contaminants and thereby increase column life, etc. These steps would take human effort to change the sample into a form in which it will be more useful than before. A sample preparation system is the solution to this problem.

It should be no surprise that there are many different types of systems for preparing samples. In the past five years, *robot oriented systems* have made great advances. The robot system has a great degree of flexibility for changing the sequence of sample transformations. Robot oriented systems use a work station approach where each work station performs the needed task while the robot moves the sample from task to task. These tasks include, but are not limited to, weighing, diluting, mixing, heating, cooling, filtering, sorbent extraction, standard addition, and sample injection. Since the work station performs the task, the preparation may be changed just by instructing the robot in a new sequence.

There are other preparation systems with less flexibility (and generally lower cost) which are dedicated to specific tasks, such as *dilution stations, sorbent extraction systems, and continuous flow sample treatment systems.* These systems may be used alone initially, and then combined with a robotic system at a later date to allow for different types of samples.

During the preparation of the sample, the chromatographic system may be *equilibrating.* Temperature control units can be stabilizing, but automated temperature control units may run through a program cycle to verify proper operation as a quality control check before the analysis.

The LC system needs to equilibrate with the mobile phase solvents. The solutions might need to have dissolved gases removed before use, to reduce the possibility of bubble formation in the lines which might cause band broadening or baseline instabilities. Solution treatment systems exist for maintaining reduced pressure over solutions in reservoirs, bubbling helium through the solution to remove dissolved gases, blanketing the solution with helium to prevent gases redissolving and using ultrasonics to remove dissolved gases. If the need for degassed solutions is critical, it might be necessary to include a monitor and alarm system to verify that the degassing system is in proper operation.

Some LC systems allow the choice of the solvent mixture for an isocratic separation or the choice of various gradients and different gradient periods through the run. These systems generally have solenoid controlled valves leading from each solvent reservoir into a low pressure mixing chamber. Under microprocessor control, the solenoids are open for varying fractions of a second to allow the chosen proportions of each solution into the mixing chamber. Over a given period of time, each solenoid valve is open for the fraction of time proportional to the chosen mixture. Gradients of any type may be created by changing the proportions with time. This type of system allows a very high degree of flexibility in the composition of the mobile phase, in addition to being easy to setup since the computer must prompt the user (or an automated controller) as to the required proportions.

Both GC and LC systems can be controlled through a data analysis system to prevent the analysis from proceeding when the background noise is too high indicating errors or equilibration failures. In addition, LC devices can include automatic valve switching systems to swap precolumns and analytical columns in and out of the mobile phase loop. These will proceed according to timed events or according to software logic which may be based on pump pressure increases, excessive noise levels detected, loss of resolution or tailing, etc. The result is a system which can detect the more common reasons for poor results and make changes in the system to attempt to correct the errors before continuing.

Sample Introduction

Once the chromatographic system is equilibrated and the samples are ready, the next step is to deliver them into the column or onto the plate. Automatic systems are also available for this purpose. For GC and LC, the most common method is to use trays to hold vials containing samples ready for analysis. Each vial is brought under a needle or the needle is brought over the vial. In either case, the needle punctures the septum and, through various means, the sample is moved into the syringe ready for injection or spotting the plate. The sample trays are indexed either sequentially or randomly according to a sequence determined by the analyst or the automation system. Autosamplers are available for TLC also; a long, flexible capillary is connected from a syringe to a mobile positioning head. The head moves over a sample vial, punctures the septum, and sample is drawn into the capillary. The head then moves over the plate and the programmed amount of sample is spotted onto the plate with a gas stream to evaporate the solvent during the spotting process. A *key* feature for laboratories which need to interrupt a sequence and run a priority sample is the ability to change the sample sequence for the remaining samples during the run.

Related to the injection sequence for GC and LC, is the initiation of the integration run by the data analyzer. This is usually achieved by the use of a pair of wires from the autosampler which closes a contact at the time of injection resulting in a low logic level being detected by the data analyzer. Two methods are available for prompting the autosampler to begin the next sample. In an open loop system, at a predetermined length of time after the injection, the autosampler begins the sequence to inject the next sample. It is the responsibility of the chemist to set the 'time out' period properly before the first sample. In the closed loop systems, the data system which collects and processes the chromatogram provides a contact closure to the autosampler telling it to begin the injection of the next sample. In more advanced systems, the data system can tell the autosampler which sample to inject next.

Chromatographic Optimization

Once the separation is underway, GC and LC systems exist to allow backflushing, column switching, and recycling for a variety of purposes. Valves to perform the switching can be controlled by timed events or from the data analysis system based on evaluation of the developing chromatogram.

On the low pressure, solvent-feed side of the HPLC pump, valves can be switched in a time-modulated fashion to modify the solvent mixture going into a low pressure mixing chamber for the generation of solvent gradients. Some HPLC pumping systems use this technique for multisolvent gradients. A microcomputer can be used to automate the on-off cycle and duration to adjust the solvent mixture as desired.

For method development, automated solvent mixture modification is aided by the use of valves allowing a wide variety of different types of columns to be automatically switched on-line to reduce mobile phase - column incompatibility problems. As the method development software chooses different mobile phase mixtures for evaluation, a column containing an appropriate stationary phase can be automatically switched in-line.

Column switching may also be used for *sample cleanup*. A portion of the effluent from a column on which many components are being eluted can be diverted onto another column for elution by another solvent. The result is a chromatogram in which many of the previously present components are not detected. Additionally, during the separation on the second column, the first column can be backflushed to remove longer eluting peaks in readiness for the next analysis.

Column switching can be used to perform *trace enrichment* of a sample. The sample is passed into a pre-column in a mobile phase which results in little movement of the sample band. Ideally, some of the other components are flushed through and switched to waste. The analytical column may then be switched in line and the concentrated compounds eluted normally.

Difficult separations may be aided by switching the effluent back into the same or similar column. The result is equivalent to using a much longer analytical column without the cost or problems associated with packing the longer column. *Recycling* in this fashion may continue until optimal resolution is obtained.

All of the options presented above can be performed manually, but are better carried out *automatically*. One approach is to prepare a time-table containing a list of the valve positions and the time to perform the switch operations. Thorough knowledge of the flow rate and normal elution times in conjunction with properly prepared columns are required. With the use of a computer controlled data system, flow characteristics, mobile phase composition, and detector signal can all be monitored to determine the optimal conditions for the switching technique to be most effective.

Detection

Automation is also available to aid in detection

of the eluted components. Most detectors for GC and LC have a signal output proportional to the parameter that they measure. Some, such as the photodiode array based detectors, are capable of providing many simultaneous outputs from different wavelengths. In addition, they may have the capability of automatically taking and sorting 'snapshots' of the full spectrum at different times during the elution of a peak. The availability of several types of information from a single separation greatly increases the flexibility of the analysis. Spectral comparisons and ratios of wavelengths can aid in the determination of the purity of a peak and allow decisions to be made by the data analysis system concerning column switching, rerunning the sample, etc.

TLC plates can be analyzed by scanning densitometers, which produce a chromatogram output by scanning a lane on the plate. Since the separated components are stationary on the plate, it can be scanned several times using different wavelengths of light, fluorescence techniques, and even chemical modification followed by the above spectral methods. Together, much information can be deduced about the composition of the sample.

Another approach to the TLC plate is the use of an image processor to digitize the image of the entire plate. Since there are no moving parts, signal averaging techniques can be used more easily to enhance the image. Once digitized, a variety of software techniques can be used to evaluate the data. The limitations include the resolution and sensitivity of the imaging system.

Fraction Collection

The separation process is beneficial for collecting fractions of the effluent for subsequent analysis by other means. Automated fraction collectors can collect the effluent at programmed intervals or can be interfaced with some detection systems to switch collection vessels when certain events are detected in the chromatogram, for example before, after, or between peaks.

Data Analysis

The *automation* of data analysis systems was a very important step in *quantitative* chromatography. Early chromatographic data collection consisted of a stripchart recording or a photograph of a plate for interpretation by a trained chemist. A trained chemist can still interpret a chromatogram, but the data analysis system available can eliminate a lot of tedious measurement and subjective evaluations from the process.

The incoming signal is most often digitized and stored for re-analysis if necessary. The digitized signal is the raw data used by the analysis system. Once in the computer there is no limit to the way that the data may be processed, analyzed and interpreted. The automation of the data analysis processes is covered in the section on computers. The results of the data analysis, however, may be used to modify the conditions for the remaining runs, for repeating runs, or documenting the performance of the system during the runs for quality control of the process and verification of proper operation.

AUTOSAMPLERS

The objective of the injection step in LC and particularly GC is to introduce the sample into the chromatographic column rapidly and smoothly enough to reduce potential extra-column dispersion. Similarly, in TLC the objective is to deposit the sample onto the plate in a small spot or smooth line horizontal to the solvent front. The advantages of automation in this stage are principally *improved accuracy, reproducibility,* and *speed.* In addition, the autosampler generally holds a rack of samples with the next sample ready for use as soon as the previous sample has been completed. This significantly increases the sample throughput.

Sample Selection

The rack and sample selection subsystems can be considered separately from the injection mechanism. The rack holds and organizes the samples for selection and injection. Samples are stored in vials, tubes, or bottles of various sizes. Some systems are required to be open to the air, others are sealed with septum caps. Heated sample racks are available for systems such as headspace samplers. With heated sample racks, a controller is necessary to maintain the temperature and verify to the data system injection mechanism that the samples have been conditioned as programmed.

Autosampler systems must ultimately access all of the samples for *unattended* operation. Many systems access each sample in the sequence in which they are placed in the rack; some allow the samples to be accessed in any order according to a sequence programmed into the data system. The correspondence between sample vial and sample identification is the responsibility of the analyst; a potential source for major error. One product uses barcoded sample labels which the vial selection mechanism reads to identify the sample and the required analytical conditions (3). Again, the chemist has the responsibility for insuring that the samples are properly labeled and that the data system is programmed with the corresponding information for each sample.

Some systems have provision for recognizing priority samples. The system injects the priority samples before continuing with the previous sequence of samples. In a quality control environment, this feature may be very important if frequent priority samples can be expected.

One feature that all autosamplers should have is a closed loop rack index method which prevents the autosampler from continuing in the event that the rack does not present the desired sample to the injection mechanism. In addition, the autosampler should alert the data system of the error to allow the system to shutdown any activities which may be adversely effected by the failure.

Once a sample vial is selected, the sample must be transferred from the vial to the column or plate. For GC and LC, the injection mechanism must also send a signal to the data system that an injection has been made so that data acquisition can begin. In most cases, a needle is used to withdraw the sample. Some systems present the vial to a stationary needle, others move the needle assembly to the vial. In either case, feedback should be available to indicate when the sample and the needle assembly are not matching properly.

LC Autosamplers

In LC systems (4,5), autosamplers generally use two methods for moving the sample into the mobile phase stream: injection valves with fixed volume sample loops, and syringe mechanisms which draw a volume of the sample into the needle and holding loop. The fixed sample loop provides a *known* sample volume and simplicity at the expense of flexibility. The syringe system allows a high degree of *flexibility* at the expense of greater complexity and potential for volumetric errors.

With the fixed sample loop systems, the sample is drawn into the loop and then the valve moves the loop into the mobile phase stream ahead of the column. The method of drawing the sample into the loop from the vial depends on the manufacturer of the equipment. Some insert two needles into the vial and blow gas into one needle forcing the sample into the other and then through the loop. Some of the sample volume is used to flush previous sample through the loop into waste. Other manufacturer's equipment pierces the top of a plastic cap which fits into the vial and pushes the cap down into the vial, forcing the sample back through the needle and sample loop.

The syringe systems use a needle, holding loop, and syringe. The needle punctures the top of the vial and the syringe draws a measured amount of sample into the holding loop. The needle and holding loop are placed inline with the mobile phase ahead of the column and the sample is flushed out onto the column. The sample is never in the syringe, reducing the possibility of cross contamination. This system allows variable amounts of sample to be measured and injected compared with the fixed sample loop methods. On the other hand, the measuring process is subject to more variations than the fixed loop method.

A provision is available to check for gas bubbles by compressing the sample in the holding loop while monitoring the pressure; failure of the pressure to increase rapidly indicates possible bubbles (or a leak in the system). This feature also has the capability of being used to verify that a sample was actually obtained; however, most systems have no provision for verification of the presence of a sample in the sample loop.

Regardless of how the sample has been loaded into the sample or holding loop, the *critical* factors affecting the development of the chromatogram are the smoothness of the merging of the loop with the mobile phase and the quality of the connections of the autosampler system to the chromatographic system. The valve which connects the loop with the mobile phase should allow a smooth merging of the two to reduce the effect of pressure variations. It should be accurate, reproducible and reliable. Similarly, the quality of the separation can be affected adversely by *improper* plumbing - the presence of dead volumes, changes in diameter and sharp corners can lead to band broadening or tailing.

Although the operation of the valves and tubing connections are most important, the method of *transferring* the sample into the loop cannot be overlooked. A change in sample viscosity could result in more time being required to flush the previous sample through the lines leading into the sample loop. If a constant flush time or volume is specified, there may be appreciable carry-over. This type of error is usually caught in the experimental design phase, when sample viscosity variations are estimated and tests run to verify that the system will still perform as desired.

A comparison of some of the options and their performance under a specific set of conditions can be seen in Table 1.

For this particular test, the syringe unit was more reproducible than the gas displacement unit or the positive displacement unit which broke the higher viscosity sample vials. The syringe unit was easier to use but had fewer control options than the other two units. The syringe unit cost more but allowed more injections to be run from a single vial. The table does not address accuracy or sample carry-over, both of which are important considerations in selecting an autosampler. With careful experimental design, proper validation, and suitable controls, most units will operate effectively. In the absence of specific requirements such as no gas connections or a particular integrator interface, the most important feature of an integrator is the ability to provide *adequate* and *reproducible* peak resolution.

TABLE 1

RESULTS OF COMPARATIVE EVALUATION OF AUTOSAMPLERS

Unit	Syringe	Gas Displacement	Positive Displacement
Precision (RSD)*			
100% MEOH	0.969%	1.68%	3.14%
50% MEOH/EG	0.490%	2.46%	broke vials
100% EG	0.417%	7.85%	broke vials
Reliability			
Overnight	yes	yes	yes
breakdowns	none	none	4
Chances of Operator Error			
	minimal	greatest	intermediate
Cost (1982			
	$8000	$5000	$7000

*Samples contained different concentrations of orthodichlorobenzene in methanol - ethylene glycol (EG) solvents

For small volume sample sizes, sample loop resolution is inadequate. A system has been described which can use any sample loop oriented autosampler (6). Between the injector and top of the column is a valve which allows the sample to flow to waste until the center of the sample plug is at the valve. The valve opens for a given period of time allowing sample onto the top of the column and then closes, purging the rest of the sample to waste. The result is an amount of sample determined only by the flow rate and the length of time that the valve was open. Sample sizes in the nanoliter range can be obtained in this way.

Some sample preparation systems have provisions for introducing samples into LC systems. Continuous flow sample preparation systems are available which prepare the sample, then allow it to flow through the sample loop of a valve injector (7). A sorbent extraction system cleans the sample and leaves it on the sorbent column. The column is placed inline with the LC column and the mobile phase then elutes it onto the analytical column (8). The system is not a general purpose autosampler; its value is determined by the need for sorbent extraction cleanup.

A general purpose autosampler is available in the form of a robotic sample preparation system (9). The flexibility of the robot manipulations allows it to use a syringe and a manual septumless injector to fill vials and place them in an autosampler rack, or to take a sample in an open test tube to a sipper arrangement which draws the sample through the loop of a valve injector. In all cases, the robot only provides the means of moving the sample from one place to another; sample introduction into the mobile phase is performed by an auto-injection system or robotic activation of a valve with a sample loop holder similar to some autosampler systems.

GC Autosampler

GC autosamplers handle liquids similarly to LC systems. The GC injection, though, must present the sample in the injection port rapidly for evaporation to occur with minimal band spreading. Rack indexing mechanisms are similar to LC samplers, but the sample must be injected rather than just moved inline. The GC injector must be capable of inserting the needle, injecting the sample, and removing the needle rapidly and reproducibly. The intercommunication capabilities must be similar to those of LC autosamplers.

There are autosamplers for specific purposes, such as headspace analysis, with special features such as thermostated sample racks. Some sampling systems use gas-tight syringes for sample introduction while others pressurize the sample vial and allow the sample to expand into the GC inlet (10,11).

Robotic sample preparation systems allow sample introduction according to existing methods to be used for GC systems with minimal changes from the manual operations (12). Robotic systems are useful for preliminary sample preparation, but are *not* currently recommended for direct injection in lieu of GC autosamplers due to speed and reproducibility constraints.

TLC Autosampler

Autosamplers for TLC systems apply samples in liquid form or pre-evaporated directly to the plate (13). There are no timing requirements as in GC or LC, but *mechanical* positioning is important. Autosamplers for liquids use sample vial racks and syringes with capillaries to withdraw the sample from the vial. The capillary is placed over the plate and the specified volume is dispensed onto the plate with a gas stream to evaporate the solvent during applications.

Autosamplers allow the chromatographer to stage a number of samples for separation and subsequently leave the system unattended. Care must be taken by the analyst to properly match the samples in the racks with their identifying notations in the data system. The autosampler should not continue to inject samples if the pump or detector is not operating; similarly, the rest of the system may need to be shut down if the autosampler has malfunctioned. The sampling systems have the capability to perform accurately and reproducibly, to minimize the time between injections, and operate overnight with a consequent increase in throughput.

COMPUTERS

The use of computers has dramatically increased the capabilities of chemical instrumentation in the past decade. Much of the advance in automation is directly due to the ability of computers to collect data, perform comparisons, and provide useful output faster and more reliably than by human means. Computers have provided the capability of automating most of the chromatographic process from conception to completion.

Data Acquisition

Probably the most *productive* first step away from totally manual operation is that of data analysis. Data analysis consists of data acquisition followed by data processing; converting the detector signal into information. The acquisition stage requires digitization of the incoming analog signal.

The rate of digitization determines the size of the narrowest peak that can be determined. Assuming that the narrowest peak is of the order of one second wide, digitizing at 20 samples per second (20 Hz) should be sufficient. Most commercial instruments digitize at rates from 20 Hz to 200 Hz. At a sampling rate of 20 samples per second, a long separation may use a lot of memory for data points. Since peaks tend to get broader as the separation lengthens, the need for fast sampling rates decreases. The system should provide the capability of *reducing* the sampling rate automatically with time or provide the capability of specifying the sample rate at predetermined times during the run. Some systems continue to sample at the same initial rate, but average or smooth several of the samples into one data point at longer time periods.

Another factor to consider is the *implementation* of the hardware and software. For one chromatographic system with one detector, most data systems will be adequate. If two detectors are used for the same separation, a data system which will handle two channels of data is necessary. If more than one chromatographic system is to be in operation at the same time, there are three approaches: dedicated, shared, and distributed.

The needs and the finances of each laboratory will dictate which approach is best. Often, a combination of the three will evolve as needs change. The type of systems chosen is less important than their ability to perform properly as intended. The following paragraphs describe each system.

In the *dedicated* approach, there is one data system for each separation unit. Each system is totally independent of all others. Dedicated equipment is ready for use as needed and will operate regardless of failures in neighboring systems. The disadvantages of dedicated systems include the cost of duplicated equipment and space constraints.

With the *shared* approach, one central system handles digitization, storage, and processing of all chromatographic data. The system must be capable of digitizing and storing data at a rate faster than all of the chromatographic systems combined are capable of providing it. This results in higher initial costs, but this will be spread over several instruments. The disadvantages of shared systems come mainly from the danger of overloading the system with more instruments than it can handle and from equipment failures which shuts down data collection for all systems simultaneously.

In the *distributed* approach, each separation system has enough dedicated equipment to continue for an appreciable length of time, but ultimately passes data or results upline to a central system for storage or further processing. This is really just a combination of the two previously mentioned approaches. The disadvantages of each are minimized. If any unit fails, the others are not affected. If the central collection unit malfunctions, the data is not lost, just held in the local unit until it is sent for. The distributed approach allows control from a central unit and totally independent data acquisition systems which adds flexibility and reduces the complexity of the central computer.

Data Processing

The *automatic processing* of data from chemical measurements has had a tremendous impact on *quantitative* analysis. The conversion of the digitized data into final results requires mathematical manipulations of various kinds. Accumulation, averaging, smoothing, transforming, integrating, and derivatizing are just a few of the purely computational operations performed. In addition, scaling, plotting, printing and displaying, with and without color, can aid in the manual interpretation of large amounts of data when a specific data evaluation process is to be performed. The options for processing chemical data provided by computer systems had resulted in widespread automation of data reduction at reasonable prices.

After the data has been digitized, the data system should store the digitized data points and then begin the process of converting the data into information. The raw data storage step is critical. Some older systems convert the raw data into peak tables by integrating the data as it is received and discarding the raw data. This method saved the cost of memory, but at the expense of repeating the separation if one of the integration parameters was incorrect.

The integration of the raw data into peak area and retention time information is the *primary* task of the data processing system. There are many algorithms for peak integration, but usually two parameters have the biggest effect on the outcome: *threshold* and *sensitivity*. Both factors relate to the interpretation of the data as peaks. Although these parameters have been discussed in Chapter 2, the subject is sufficiently important for it to warrant further treatment. The threshold level indicates the level at which an increase in the baseline signal may indicate the start of a peak; below the threshold level it is probably not a peak. The sensitivity factor is related to the rate of increase in an increasing signal. If the rate of increase is too low, it is probably baseline drift and not a peak. But if the rate of increase exceeds the sensitivity and the threshold level is passed, the signal probably indicates a peak.

For unresolved and tailing peaks, the algorithm chooses to use baseline corrections based on the initial threshold and sensitivity levels. These decisions may result in some peak areas being calculated one way in one chromatogram and another way in a second chromatogram. Since the area calculation is based on arbitrary threshold and sensitivity levels, reintegration of both chromatograms with new levels may result in the same type of area computation being used for the same peak in both chromatograms. The result is a more valid integration using the stored data without the need for rerunning the sample.

Once peak areas or heights have been calculated

and stored, the concentrations or other information of interest may be computed as desired. In particular, the data system has the ability to compare the results with those from previous analyses of the same sample, related samples, and standards.

Post Processing

Another aspect of the utility of storing the raw data is the ability to *recalculate* the report using new parameters, sample weights, etc. If new data comes to light concerning the sample, the data may be again reanalyzed as if the separation had just taken place. It has to be emphasized that this ability will not correct a poor or faulty separation; it just allows a distinction to be made between the quality of the separation and the quality of the analysis of the data.

Control

During the chromatographic separation, a variety of activities may be controlled by the data system at preset intervals. These non-interactive control activities are called 'timed events'. The data system watches an internal clock and initiates actions at the pre-selected time.

Sampling rates may be changed during the run, columns may be switched at selected times, the gain of the digitizer may be changed to allow digitization of larger or smaller peaks than the rest of the chromatogram; temperature changes in GC and gradient changes in LC may also be initiated at preset times. Although this ability can be very important to an automated separation, the events are programmed by the *analyst* based on prior knowledge of the time that an event needs to occur. Unexpected changes in the nature of the separation may lead to loss of synchronization between the timed events and result in unusable data.

Quantitative analysis is particularly vulnerable to the activation of timed events at *inappropriate* stages of the chromatographic separation. Examples involve changes in operating parameters during critical phases of the separation; initiation of changes in solvent gradient, thermal ramping and digitization rates. Stable operating conditions and well known chemical systems are required for effective use of timed events.

Interactive Control

The solution to the loss of synchronization is to monitor the progress of the separation and perform the action in synchronization with an event in the separation. For example, if the fourth peak in the separation is much larger or smaller than the rest, the gain may be changed after the completion of the third peak and before the fourth. Temperature or gradient programs may be started at suitable times determined by monitoring the separation in real time. The lifetime of a column may be increased by monitoring the signal until the solvent peak passes and then switching the analytical column inline.

All of the activities just described are expected to happen in real-time; the data system has to monitor the digitized signal from the detector, apply the integration routine to the data, recognize an event, and signal the hardware to perform the action. Interactive control does not always have to be real-time; the results of the analysis may be used to determine the next separation to take place.

One of the greatest advantages of a flexible data system, is the opportunity to change the sequence of analyses based on the results from previous analyses. An unexpected peak may have been out of range of the digitizer; the sample may be scheduled for re-analysis with a different digitizer gain. Evaluation of the retention times or peak shapes over several chromatograms may indicate gradual degradation of the quality of the separation. The result may indicate the need to switch to a new analytical column , allow it to equilibrate, run several standards to verify proper operation, possibly rerun the last couple of samples, and then continue with the remaining samples.

The following example demonstrates a separation in which interactive control can automate not only the analysis itself, but also the steps necessary to insure that the separation has been conducted properly. The compendious method specifies the measurable requirements for an effective separation.

The USP XXI monograph for Cefazolin (14) calls for the use of liquid chromatography and specifies that the separation meets the following requirements:

1. Column efficiency not less than 1500 theoretical plates, tailing factor not more than 1.5,

2. Resolution between analyte peak and internal standard peak not less than 4.0,

3. Relative standard deviation between replicate injections not more than 2.0%.

For the separation to be considered acceptable, the above criteria must be met. The flexible data system would allow modifications to the equipment operating conditions (e.g, flow rate or mobile phase composition) to bring the separation back into control. Interactive control also includes printing or setting alarms and awaiting operator intervention before continuing. An example would include pump failure. The system should recognize a failure condition and suspend operation rather than continue to inject a sample, wait the allotted time, integrate the flat baseline, and continue with the next sample.

The computer is ideally suited for interactively controlling all aspects of the chromatographic separation. The cost of microcomputer chips is low enough that some instruments use more than one to control various internal functions. In addition, chemists have demanded that new equipment has interfaces for control by external computer systems. The external interface should allow complete and total control to the extent that the analyst should be able to provide all operation commands without having to touch the instrument. Of course, the instrument should also be able to be run entirely as a stand-alone system without the need of an external controller. This allows it to be in full, but possibly manual, operation in the event that the external controller is out of service.

The degree of automation required for a chromatographic separation should be considered when specifying the equipment to be used. The instruments chosen for use should at least have the essential features needed. For example, an LC detector may appear to be fully automated, but if automatic wavelength changes are required and not available, an analyst would have to attend to the 'automated' analysis just to change the wavelength when required. The extra cost of the interface for external control will be justified if the equipment is planned for use in an automated environment.

For automation of existing manual equipment, mechanical modifications or other approaches, such as robotics, may be necessary. If the manual step requires keyboard interaction, the keyboard may be by passed with a custom keypad emulation interface (15). Under computer control, the interface would electrically connect the wires necessary to make the instrument operate as if a human were pressing the keys.

Interactive control is best demonstrated with some of the chromatography method development systems (16-20). The method development system attempts to determine the conditions for the best separation possible under isocratic development with up to four solvents. The system uses resolution criteria to evaluate each separation in conjunction with previous separations and determine the mobile phase mixture for the next separation. The software includes algorithms designed to recognize improved separations and direct the mobile phase choice in the direction most likely to result in further improvement. Once the system is started, it can be left unattended while the optimization proceeds.

Interactive Problem Solving

Many data systems display the data on a monitor in graphic form, some allow the user to interact with the data on the display. This feature can effectively increase the speed in which the chemist comprehends and improves the quality of the information available (21). The system allows the human faster and better access to the information than before.

Chromatographic integration algorithms are created to handle as many of the different types of baseline, peak shape and over-lapping peak situations as possible. The algorithm uses threshold and sensitivity factors to discriminate between noise and information, but the initial setting of the parameters is usually based upon *experience*; they may not necessarily apply to any particular separation. The ability to post-process the chromatogram with new parameter choices allows effective use of the raw data without unnecessary re-analysis. Interactive software can aid in the determination of appropriate parameters by letting the chemist rapidly see the result of parameter changes.

Assume that the chromatogram is displayed on the screen. The default parameters show the integration baseline under a peak as starting too soon due to the noise level just before the peak. The chemist moves a cursor, pushes a couple of buttons, and the small area expands to fill the screen. The chemist recognizes the error and types new threshold and sensitivity parameters. The screen is erased, and the data is re-drawn with the new integration baseline using the modified parameters. The chemist again examines the area in detail, is satisfied with the result, ends the transaction. A hardcopy of the chromatogram, baseline and parameter choices included on the page, is printed for documentation purposes. The raw data is not changed, the parameters are modified to better suit the particular separation.

Other interactive applications include chromatogram comparisons and equipment setup prompting. For comparison purposes, seeing several chromatograms superimposed in pseudo-three dimensional format, possibly in color, affords the analyst the opportunity to quickly ascertain similarities and differences between separations.

Many automated data systems now have interactive menus for setting instrument parameters. The menus usually give the range of choices and suggest default settings to reduce errors in the setup. Some of the newer systems use highly graphically oriented screens, with point devices, which reduce the length of time that a new user needs to begin performing effectively. The graphic orientation aids the experienced user by allowing the status of the system to be determined more quickly.

Archiving

The use of a computer for a data system allows

many options for data storage. The usual include numbers on paper, chromatograms on printers or plotters, raw data on tape, floppy disks, and hard disks, or it may be transmitted to another computer for further analysis or storage.

It may be desirable to store the raw data, intermediate data (peak tables and compound identifications) and final data (component concentrations), calibration data, system setup, etc. The data may be stored for limited periods of time for use in longer term studies and for possible reprocessing if needed. Reuse of the data in other studies would be much easier if available online.

Compact disk data storage is now gaining momentum. Currently affordable technology provides read-only capability. Thus, large data bases of information are the most practical applications for CD-ROMS (Compact Disk Read Only Memory). One 4.7-inch plastic disk can hold up to 600 million bytes of information (22). Ultimately, users may write data to CD-ROMS instead of microfilming for archiving purposes, allowing data system access to older records.

Communications

A big advantage of computer-based data systems is communications. The raw data might be transferred to a larger computer for more sophisticated analyses, or just to offload processing from the smaller machine. The smaller system might send a request to a larger system for the specifications of a sample, compute the results obtained, and send the finished result back to the larger system for storage in a finished data base. Another possibility is for a central system to store all method parameters for each method. The distributed systems would then all be down loading the exact same procedure; and any changes to the procedure at the central system level would immediately be available to all of the remote systems.

Communications capabilities also allow systems to be created, in which each remote unit sends the central system the status of the equipment and the progress of the analysis. Everyone with access to the larger machine would be capable of seeing the operating conditions and progress of samples in the laboratory without bothering laboratory personnel or interfering with the analysis in progress.

A 'modem' is a device which converts a computer's digital signals into analog signals which can be transmitted over the telephone system. Collaborative studies would be facilitated by allowing all researchers to *share* data easily and rapidly via modem and telephone between systems. Rapid international collaboration is possible via modems through the commercial global networks. A local call to the network service via a modem can have data sent through a satellite link to any city the network serves worldwide.

Artificial Intelligence

Artificial intelligence in the form of "expert systems" is now currently available in one form or another for most computer systems. The expert system software which ran effectively only on larger computers in the early 1980's is available on professional microcomputers.

An 'expert system' is a computer program which contains the knowledge of one or more human experts in the form of a 'knowledge base'. The knowledge base is a set of rules which apply to a group of problems within a specific scope. A common approach is the use of IF-THEN rules to form the knowledge base (23,24). An 'inference engine' is a portion of the program which takes the known information about a problem and applies the rules from the knowledge base to generate a list of the most probable answers. A 'user interface' (25) is the portion of the program which handles interactions with the user. The final result of using an expert system is a list of the most probable answers along with the probability of their correctness. Additional output available include the sequence of rules that are employed to evaluate the data and to arrive at the answer, forming a logic sequence which can also be evaluated manually for educational purposes and for program performance verification.

For example, an expert system might be used to determine the best column and initial conditions for the development of a method (26). Through an interactive session, the system would obtain basic information on the solutes such as molecular weight, functional groups, etc., and then sort through the knowledge base for rules which fit the known data. As rules are evaluated, the session progress is displayed for verification. An answer is found if all of the applicable rules lead to a conclusion; if not, the knowledge base needs to be updated with rules to cover the new situation. In this fashion, either answers are obtained or the knowledge base is updated for future encounters.

The use of the expert system thus allows a result to be obtained with full use of the knowledge stored from previous evaluations in a fashion similar to that performed by the human expert. Quantitative chromatography can benefit from the use of expert systems through applications ranging from determination of the best starting conditions for a separation, to trouble shooting equipment malfunctions. The use of expert systems depends on the *development* of knowledge bases for the types of problems expected.

Retention Indices

The development of retention index libraries for chromatography is promising from two standpoints: identification of unknowns and methods development. A GC retention index library is available (27,28). The library is used to provide a data base of columns, packings and conditions for a desired separation using capillary GC and is indexed alphabetically or by molecular formula. The goal is to allow identification of an unknown through the use of GC only. Indices are currently used for determining initial conditions for the separation of standard compounds.

The use of retention indices in conjunction with expert systems and the interactive control of instrumentation are powerful automation tools made possible through the availability of computers in the laboratory.

ROBOTICS

In 1982, the first question asked of a chemist using a laboratory robot was usually, "Does it make drinks?" Since then the laboratory world has taken robots seriously and there are now several manufacturers offering robots for a variety of purposes (29).

Most of a chemist's manual activities take place within a couple of cubic yards on a benchtop. A robot that could duplicate these activities would only need to reach within that space and would be directly usable since the activities are generally *well* defined. Consider a typical sample preparation procedure: an opaque liquid sample in a test tube is weighed, an aliquot of a standard solution is added to a similar test tube and a similar volume of a solvent is added to a third tube, an aliquot of an internal standard is added to all test tubes. All three are placed on a mixer for ten seconds, all three are allowed to stand for 60 seconds, several milliliters of each is drawn into a syringe through a filter, the filter is discarded, the contents of each syringe is dispensed into a separate vial, the vials are capped and placed in an autosampler tray, a button is pressed to initiate an LC analysis of the sample, the standard, and the blank.

Each step of the sample preparation is well defined, each operation is easily within reach, each operation can be performed by a robot. Place several samples in tubes within reach of the robot, tell it how many are there, and go do something else.

Since most of the activities take place in a relatively small space, a robot sitting on the table should be able to use whatever was within reach. Robots currently used for laboratory functions are tabletop units with a reach of about a yard and an accuracy of about a millimeter. This gives it the ability to pick up a test tube and carry it across the benchtop. Current technology does not allow the robot to have sight or touch, they must rely on everything being in the place where it is supposed to be. For that purpose, the robot is bolted to the table as are mixers, test tube racks, filter holding racks, liquid dispensing nozzles, etc. To fill a test tube, the robot will move to the coordinates that are supposed to be under the nozzle of the dispenser and send a signal to the dispenser to begin the task. An enhancement to this system might be a microswitch near the dispenser so that the robot controller could tell the arm to move the test tube to the coordinates beneath the nozzle, the controller could then check the microswitch to determine if the test tube closed the contact. If the contact is closed, begin dispensing; if not, the tube is not there, do something else. In this manner, each step in the sample preparation could be defined to the controller so that the entire sequence, error checks, etc., can run automatically.

These tabletop robots are capable of performing many of the same functions that a human arm, hand, and fingers can. The units available today are based on three coordinate systems: cartesian, cylindrical, and revolute (30). The *cartesian* system uses two horizontal arms and a vertically moving hand in an arrangement similar to an X-Y plotter mechanism.

The *cylindrical* system has a vertical pole attached to the base. The pole has an arm that can ride up or down the pole and extend in or out. The pole can rotate 360 degrees. The arm has a hand that can rotate 360 degrees and the hand has two fingers that can open or close. The space that can be reached is a hollow cylinder; around, up and down, out and in (but not all the way in to the pole).

The *revolute* coordinate system is more complicated. The base may rotate 360 degrees, the arm may raise from horizontal to vertical, the forearm may swing from in to out, the hand may wave and rotate, and the fingers may open and close. The result is a hollow hemispherical space that can be reached. The cylindrical coordinate system is easier to conceptualize than the revolute system, making it somewhat more easy to use, but the revolute system allows a robot to reach over or around objects that the cylindrically oriented robot cannot reach. Careful placement of all objects designed to be reached is a requirement for full utility of either system.

One of the robot systems uses multiple hands. There are hands with fingers for test tubes and vials, hands with syringes, and hands with tubing connected to liquid dispensing systems. Each hand is "parked" in a hand station. The arm moves the wrist in front of the hand and extends the wrist until the hand is in place. Changing the hand is a matter of reversing the process.

Complex programming is not necessary to teach the robot what to do. The robots are controlled by a dedicated computer which stores coordinate locations and names for the locations. The robots are taught the locations of the objects on the table and the positions of the hands needed prior to grasping an object by moving them into the desired position and giving the position a name. The names and associated coordinates are stored into a table. Sequences of movements are created by making lists of the names. For example, to grasp a test tube and weigh it, the sequence might be:

OVER.TUBE
OPEN.FINGERS
DOWN.TO.TUBE
CLOSE.FINGERS
OVER.TUBE
OVER.BALANCE
ON.BALANCE
OPEN.FINGERS
GET.WEIGHT
CLOSE.FINGERS
OVER.BALANCE
OVER.TUBE
DOWN.TO.TUBE
OPEN.FINGERS
OVER.TUBE

The whole list of positions might be called *weigh.tube.* Anytime the command was *weigh.tube*, the controller would send the robot to each location in the list and the result would be a weight stored in the controller. Objects like balances, mixers, heaters, dispensers, etc. are called stations. The robot arm itself is a station. The list that the controller keeps includes positions for the syringe in a dispenser, the valve from the solvent into the syringe, the speed of a mixer, or the tare of a balance. The list of steps includes robot positions and dispenser valve and syringe positions as well. The method is one small list of other lists, those lists may have positions or other lists also.

For example, a list called *add.standard* may have *weigh.tube* and *dispense.standard.into.tube* in it. When *add.standard* is called, it calls *weigh.tube* and then *dispense.standard.into.tube*. When the list has been completed, the method is completed (31).

There are three approaches to the sequence in which robots can perform their activities: batch, sequential, and concurrent. In *batch* mode, the robot performs one activity to all tubes before proceeding to the next activity. Operations which must be performed as rapidly as possible benefit most from batch processing. For example, weighing samples which are dissolved in volatile liquids might require that the initial weight of each solution be recorded as rapidly as possible. The robot would move each tube to the balance as soon as the previous one is returned. Thus, each tube would be weighed as soon as practical.

Since all samples are carried through each step before the next step is begun, no samples are completed until all are waiting for the final step. Thus, the finished samples are not available until the last step of the sample preparation has been reached. In *sequential* mode, the robotic system performs all activities on one sample proceeding to the next. Each sample is available as it is prepared before the next sample is begun. Finished samples are available throughout the sample processing period. A major advantage of sequential processing is the timing of operations; each operation is performed for each sample in the specified sequence. Since the sequences are identical, even events such as sample transportation (usually) take equivalent times.

Sequential operations may be important when kinetic effects can occur. For example, compare sequential and batch modes during the addition of a reagent and its subsequent deactivation. Assume that the deactivation step takes longer than the addition step. During the sequential processing, each sample undergoes reagent addition and then deactivation begins a reproducible time after addition for each sample. During the batch mode processing, however, after the addition of reagent to each sample, deactivation begins with the first sample. Since the deactivation step takes longer than the addition step, by the time the sequence reaches the last sample, it has been exposed to the reagent for a longer period of time than the earlier samples. The third approach, *concurrent* mode processing, may complicate timing considerations even more.

For procedures which consist of several stages, neither batch nor sequential mode processing may make full use of the equipment available during processing times. For example, during mixing operations, the weighing equipment may be inactive. Concurrent mode processing attempts to improve throughput and equipment usage by interleaving the processing of each samples as it proceeds through the various stages.

An example illustrates the complexities involved; assume ten samples need to be weighed, heated and mixed. Assume also that transportation takes 5 seconds, weighing takes 30 seconds, heating takes 180 seconds, and mixing takes 60 seconds. Additionally, the heater has room for four samples. The computer system controlling the robot and each operation station must keep track of which sample is currently at which operation, how long it has been there, which sample to move next, where it is going, etc. In addition, certain operations, such as heating, may require precise timing for removal.

Scheduling the sequence and timing will be a critical task to avoid situtations where conflicting events are required to be performed at the same time with the same equipment.

A complicating factor in the determination of appropriate timing and event sequences is error detection and recovery. The performance of critical steps in robot assisted procedures should be verified through appropriate means. For example, the removal of a test tube from a balance should be verified by using the test tube to press a button or by checking the balance after removal to verify that the weight reading returned to the empty weight. If the verification has indicated that the tube is still on the balance, the removal step should be repeated. When the verification indicates that the tube has been removed, then the procedure can continue at the next step. Changes in the sequence may be necessary to correct for the time lost in removing the tube from the balance.

Batch mode is valuable when each sample needs to reach a certain stage before proceeding to the next. Sequential mode is important when reproducible timing is necessary. Concurrent mode provides the fastest overall sample throughput. The *constraints* of each chromatographic method will dictate the mode or combination of modes which will provide the best system operation.

The advantages of robotic-based automation is *flexibility*. The robot can manually perform operations that were initially manual. Whole new approaches to automate manual methods are not necessary. Since the robot is taught through lists, new or modified activities simply require changing the list. The robot can load an autosampler or perform the injection manually with more reproducibility than a person. The robot system can run 24 hours per day. The controller can display the list and steps performed as they are encountered; the same list may be printed as confirmation that each step did occur, including the weights obtained and any switches that the robot may have closed while performing the list. Such a list can be used for *validation* of the method and *quality control.*

The robot based system is not necessarily faster than a person; it is tireless, though, and can perform continually. It may require *more* maintenance than other sample preparation systems; the trade-off is flexibility. Expect to have a chemist knowledgeable in the method available for maintenance, modifications, and validation full time for the first six months of use and several hours a month thereafter.

Although the robot system is primarily advantageous as a sample preparation system, any manual operation within its reach may be automated (32). This includes pressing keys on keyboards, handling fraction collection vessels, injecting samples, moving TLC plates, etc. The weight that the robot can lift depends upon the make and cost of the robot,but is generally in the 1-15 pound range. Examples of robot-assisted methods include sample preparations of pharmaceuticals (9,33), direct robotic injection of samples in kinetic studies (12), studies in organic synthesis (34), and 18 papers presented at the 1985 Pittsburgh Conference on the topic of laboratory robotics.

The information in Table 2 includes results from several papers found in Volume 1 (1984) and Volume 2 (1985) of "Advances in Laboratory Automation Robotics" Zymark Corporation, Inc. The results included are presented only as a comparison of the difference in *accuracy* and *precision* between *manual* and *robotic* sample preparation. The final quantitation method was either GC or LC for the listed results. In all cases, the robotic system served only to prepare the samples and transfer them into autosamplers or into autosampler vials; the actual sample introduction into the chromatographic system was via an autosampler or manual injection valve rather than the robotic system directly moving the sample into the mobile phase stream.

TABLE 2

Reference		Manual	Robotic
1 113	Tridiphane Recovery in Rodent Chow		
Recovery		93%	86.5%
Rel. Precision @ 95% confidence level		4.5%	3.77%
1 120	Amount of Sucrose in Food		
Amount found		44.30%	43.94%
% Coefficient of Variation (CV)		5.24%	6.13%
1 120	Amount of Lactose in Food		
Amount found		8.01%	8.10%
% CV		7.34%	8.94%
1 120	Amount of Theobromine in Food		
Amount found		2.54%	2.46%
% CV		2.43%	3.47%

The difference between the robotically prepared samples and the manually prepared samples is directly related to the ability of the system implementors to convert their *expertise* in analytical chemistry into a program for the sample preparation system. Since the manual and robotic systems both used the same chromatographic systems and injection mechanisms, differences between the robotic and manual results were due solely to the sample preparation prior for the chromatographic step. The data indicate the importance of the entire treatment of the sample for the quality of the final result.

THE ROLE OF THE CHEMIST IN THE AUTOMATED LABORATORY

Throughout this chapter the discussions have focused on how automation can do what the chemist already knows how to do. Equipment cannot be designed, constructed, and implemented to automate a task if that task is unknown. The principal advantage of automation is just that; converting a task that required human attention into one that does not. The chromatographer is changing his or her perspective from performing the task to creating and running a system which performs the task as well as, if not better, than before.

The time made available and the change in perspective allow the chromatographer to focus on 'problems' rather than tasks. The problems were there all along, there was no time to address them before. Once a problem is solved, if encountered again, it becomes a task. The chemist can concentrate on what he or she has been all along - a problem solver.

A portion of the chromatographer's problem solving will be devoted to maintenance, diagnosing, and arranging for repairs of the task performing system. Validation of the automation system will provide a baseline of performance information used for maintenance and calibration of the system, as well as demonstrating the effectiveness of the system.

Change is a constant part of life, and it will have an impact on the operation of any automatic system. Modifications to systems will be required, and thorough understanding of the existing systems will be essential to insure that changes do not effect the validity of results.

The chemist will increasingly have the opportunity to plan for automation in light of the goals of the organization. Optimization of laboratory systems has to coincide with those goals to be effective. New and modified functions must be evaluated for both performance and cost and implemented accordingly.

SYNOPSIS

Automation is the process of *turning routine tasks* over *to systems* for performing those tasks. The task may be as simple as starting an integrator while injecting a sample into a GC by having the syringe press a switch. It can be complicated as a complete system to determine appropriate columns and conditions for a separation, including optimization of the separation; perform sample cleanup; run the samples and standards; archive the results and transmit them and the method to a sister facility halfway around the world.

The *need for automated analysis falls* into *four categories*:

1. *Time* - a vast amount of time is spent by chemists preparing samples and equipment and attending to the analysis. Additionally, the total time required from sample receipt to final result and the time necessary to complete testing on all samples have an impact on lab staffing and possibly the performance of systems requiring those results.

2. *Reproducibility* - sample preparation and introduction into the chromatographic system play an important part in the quality of the analysis.

Successfully implemented automation systems can improve the reproducibility and confidence level of the results.

3. *Reliability* - automated system improves the reliability of the steps performed, also increases confidence in the quality of the results. Error analysis is aided by the reliability of automated systems since the errors can be expected to occur consistently under the same conditions.

4. *Cost* - cost savings are often the *most important need for automated analysis*; it is the numeric justification for the benefits described in the three preceding paragraphs.

The *importance* of chromatographic *automation* can be seen in laboratories ranging from *production control to medical research* and from *forensic chemistry* to *environmental monitoring*. Automation can aid sample preparation by performing all of the steps necessary to prepare the sample for the analysis, reproducibly and reliably. It can improve the control of equilibration conditions to reduce variability in the results. Sample introduction systems trace and maintain the samples and introduce them into the chromatographic system. The *optimization of the separation* via the use of gradients, column switching, backflushing, etc. *has advanced dramatically with the aid of automation.* Detection system have benefited from electronic improvements leading to simultaneous multi-wavelength spectral data acquisition and the combination of detection systems providing better information than each technique individually. Automation improves the reliability of fraction collection through the use of features in the development of chromatograms as well as collection at predetermined timed intervals. *Data analysis* has gained *tremendously from automation*; many of the subjective aspects of chromatographic analysis have been eliminated. Additionally, the *speed of data processing* and the *volume of data evaluated* during that time *had increased.* Autosampling systems provide two functions: sample presentation and sample introduction into the chromatographic system. The presentation function allows staging samples for analysis and presenting the proper samples and standards in the desired sequence. The operation of the presentation system should be tied to the rest of the chromatographic system such that a failure in either part does not compromise the operation of the other or adversely affect the pending analyses other than by lost time. LC sample introduction systems use *two types of sample measuring devices: fixed volume holding loops*, and *syringe filled holding loops.* The sample is introduced by moving the holding loop into the mobile phase flow. Various types of equipment are available, but the important feature is the ability of the device to introduce the sample without compromising the quality of the separation. *GC systems require rapid and smooth introduction of the sample* into the mobile phase stream. Specific purpose injectors are available for techniques such as headspace analysis. *With TLC autosamplers* there are no timing requirements as with LC and GC but *precise mechanical positioning for sample placement is required. Computers have made most of the automation advances possible.* They handle the range of functions from instrumental control to data acquisition systems and from digitization of the raw analog signal to communication of results to databases globally. The digitization of data can be handled for systems from a centralized computer or from multiple distributed systems. While intermediate configurations vary, distributed data acquisition system generally provide more flexibility, portability and minimizing redundancy in the event of system failures. The storage of the digitized data allows for a wide variety of data processing and post-reprocessing options not available before. In addition, data may be compared across many runs and used to provide interactive control of an analysis by varying the conditions during the separation. The *computer affords the analyst a rich environment* for the processing of the data. Possible options and evaluations may be investigated interactively even to the point of simulating the separation under a variety of conditions. The data and results may be archived for future reference of communicated globally to others. *Software in the form of 'expert systems' is now available to evaluate data bases of knowledge* and infer solutions to problems where absolute answers are unavailable. *Robotic systems* are in use to *reduce* the *human costs and errors* associated with manual movement of materials between the systems. Robots are available in several configurations and have a variety of hands for manipulating objects. The robots are taught by example or simple languages and can perform tasks serially, in batches, or concurrently in an optimized fashion. *The complexity of the programming goes up dramatically in concurrent-mode handling. Robotic systems* are *not generally faster* than humans, but they *are tireless* and can *work in hazardous environments.* Their major advantage is flexibility in the performance of tasks and the ease of reprogramming the operations. The *current literature indicates* that the *precision and accuracy of robotic systems are still quite variable* compared with manual operations. The chemist faces a changing laboratory environment with the implementation of automation devices. The *automation systems* will *reduce* the amount of *repetitious work,* freeing the chemist for the unique and challenging types of tasks which humans do best. Every aspect of the chromatographer's function which can be automated will result in a saving of time, effort, reduced errors, and money. *The biggest impacts on a totally manual method are autosamplers, robots,* data systems, and the *computers,* large and small, which make them *possible.* Is there a need for automated analysis? Yes, in one form or another, automation will aid in allowing the most expensive part of the laboratory, the people, to perform the functions they are best suited for.

References

1. H.R. Felton, Am. Lab., 5 (1980) 105.

2. A. Stevens and T.L. Brooks, Am. Lab., 10 (1985) 22.

3. D. Gillen, K. Haak and G. Ouchi, Am. Lab., 5 (1981) 94.

4. P. Winkelbauer, Am. Lab., 5 (1982) 44.

5. E. Walcek and D. Ball, Am. Lab., 9 (1982) 118.

6. V.L. McGuffin and M. Novotny, Anal. Chem., 55 (1983) 580.

7. D.A. Burns, J.I. Fernandez, J.R. Gant and A.L. Pietrantonio, Am. Lab., 10 (1979) 79.

8. L. Yago, Am. Lab., 10 (1985) 118.

9. J.H. Johnson, R. Srinivas and T.J. Kinzelman, Am. Lab., 9 (1985) 50.

10. E. Jones, M. Davis, R. Gibson, B. Todd, and R. Wallen, Am. Lab., 8 (1984) 74.

11. P. Gagliardi, G.R. Verga and F. Munari, Am. Lab., 5 (1981) 82.

12. C.H. Lochmuller, K.R. Lung, and T.L. Lloyd, Zymark Newslett., 2 (1985) 6.

13. D. Rogers, Am. Lab., 5 (1985) 15.

14. United States Pharmacopeia, Vol XXI, pp. 174.

15. P.M. Wiegand and S.R. Crouch, Talanta, 32, (1985) 37.

16. J.C. Berridge, Chromatographia, 16 (1982) 172.

17. J.J. Kirkland and J.L. Glajch, J. Chromatogr., 238 (1982) 269.

18. J.C. Berridge, J. Chromatogr., 202 (1980) 469.

19. R. Lehrer, Am. Lab., 10 (1981) 113.

20. J.L. Glajch and J.J. Kirkland, Anal. Chem., 55 (1983) 319A.

21. P. Batchelder and G. Lawler, Am. Lab., 9 (1985) 32.

22. Chemical & Engineering News, 63 (1985) 8.

23. R.O. Duda and J.G. Gaschnig, Byte, 6 (1981) 238.

24. R.H. Michaelsen, D. Michie and A. Boulanger, Byte, 10 (1985) 303.

25. R. E. Dessey, Anal. Chem., 56 (1984) 1200A.

26. R.E. Dessey, Anal. Chem., 56 (1984) 1312A.

27. J.F. Sprouse and A. Varano, Am. Lab., 9 (1984) 54.

28. H.M. McNair, M.W. Ogden and J.L. Hensley, Am. Lab., 8 (1985) 15.

29. S.A. Borman, Anal. Chem., 57 (1985) 651A.

30. R. Dessey, Anal. Chem., 55 (1983) 1100A.

31. R. Dessey, Anal. Chem., 55 (1983) 1232A.

32. G.L. Hawk, J.N. Little and F.H. Zenie, Am. Lab., 6 (1982) 96.

33. J.G. Habarta, C. Hatfield and S.J. Romano, Am. Lab., 10 (1985) 42.

34. G.W. Kramer and P. L. Fuchs, Byte, 11 (1986) 263.

Quantitative Analysis using
Chromatographic Techniques
Edited by Elena Katz

Chapter 9

PHYSICO-CHEMICAL INFORMATION FROM PEAK SHAPE AND WIDTH IN LIQUID CHROMATOGRAPHY

Eli Grushka and Shulamit Levin

INTRODUCTION

Chromatography, in all of its variations, is a well known technique which can yield information not only about the purity of a certain compound, but also about many of *its physico-chemical* properties. The chromatographic process depends on the thermodynamic and kinetic properties of the system, consequently, these properties can be measured by, and calculated from, chromatographic data, provided that the data is *precise* and *accurate*. At least two monographs (1,2) and many reviews (*viz.* 3-5) have been devoted solely to the subject of obtaining physico-chemical parameters from the chromatographic data.

In general, the majority of the measurements involves accurate determination of retention times, and their conversion into the appropriate physico-chemical properties. In this manner, retention data can be used to obtain solubilities of gases in liquids (6), activity coefficients (7), virial coefficients (8), complex formation constants (9), enthalpies of adsorption (10) and other linear free energy parameters. Since all these parameters affect the overall solute partition coefficient (K), they can be determined by careful measurements of the retention time, which is directly related to K. Gas chromatography, in particular, proved to be a fertile area for such measurements. However, utilization of liquid chromatography for the same purpose is also becoming more common.

While retention data is routinely utilized in chromatography, relatively little is described in the literature concerning the use of *peak areas or peak widths* for the calculation of physico-chemical parameters. For example, in the case of gas-solid chromatography, a modest amount of work can be found relating peak shapes to transport and kinetic parameters (11). This type of information is of great interest to the chemical engineers, and indeed they have done most of the research on the topic, utilizing primarily gas chromatography.

Although this chapter will concentrate on the information that can be obtained from peak areas and widths, as measured in liquid chromatography, pertinent references describing the utilization of gas chromatography for the same purpose are also provided. Two aspects of peak width and area will be considered: firstly, the determination of *liquid diffusion coefficients* using the "chromatographic broadening technique" and secondly, the determination of *adsorption isotherms* in liquid chromatography. It should be noted that the use of liquid chromatography to obtain peak areas of a reaction mixture as a function of the reaction time, in order to calculate formation constants and other kinetics-related quantities, is not included in this chapter. The reason for this exclusion lies in the fact that, in this case, chromatography is used to obtain data which, in turn, will be applied for the calculation of parameters not related to the actual chromatographic process. In other words, chromatography is used as an ancillary technique and not as the *raison d'etre* of the experiment. Many reports can be found in the literature utilizing liquid chromatography for this purpose (5, 12, 13).

DETERMINATION OF LIQUID DIFFUSION COEFFICIENTS BY CHROMATOGRAPHIC BROADENING TECHNIQUES

Theoretical Approach

A chromatographic column of any type acts, at

least in theory, as a Gaussian function generator. Thus, a sample zone, injected as a very narrow plug (a delta function) at the head of the column, will emerge as a broad, Gaussian-shaped peak. In a packed column there are a number of processes contributing to the band broadening phenomenon, and they include molecular diffusion in both the stationary and mobile phases, packing inhomogeneities, slow mass transfer between, and in, the stationary and mobile phases, etc. In such cases, the dispersion of the chromatographic peak is a complicated function of many parameters, however, when the column is empty tubing, the situation is much simpler, and much better understood. Taylor (14), Aris (15), and Golay (16) were among the many who developed analytical solutions for the mass balance equations, describing the profile of the sample zone migrating down an empty tube under laminar flow conditions. Using their results, the second central moment, or the variance (σ^2) of the solute's concentration profile can be written as:

$$\sigma^2 = 2D_m l/u + r_t^2 ul/24D_m \qquad (1)$$

where D_m is the diffusion coefficient of the solute in the mobile phase,

u is the mobile phase velocity averaged over the tube cross section,

r_t is the tube radius

and l is the tube length.

All the quantities in equation (1), with the exception of D_m, are either system constants, such as l and r_t, or can be determined experimentally, such as u and σ. The diffusion coefficient, therefore, can be easily calculated from equation (1). From a practical point of view, it is simpler to measure the ratio (σ^2/l), which is the definition of the chromatographic plate height (H). Equation (1) can be rearranged to provide an explicit expression for the diffusion coefficient:

$$D_m = 0.25u[H \pm (H^2 - r_t^2/3)^{1/2}] \quad (2)$$

This equation, being quadratic, has two roots, of which only one is physically significant. A value of the mobile phase velocity determines whether the negative or positive root is the solution of interest. From Figure 1 it is seen that at low velocities the positive root is the desired solution, while at high velocities the negative root is the significant one.

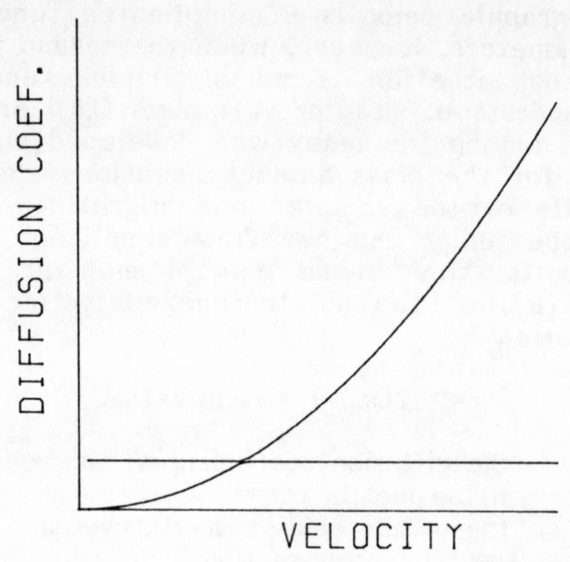

Figure 1. The roots of equation (2) as a function of the mobile phase velocity.

The change in the sign of the real solution occurs at the minimum of the *HETP* vs. *u* curve. At that point, *H* is equal to $r_t/(3)^{1/2}$ and equation (1) has only one solution:

$$D_m = u_{opt}\ r_t/(48)^{1/2} \quad (3)$$

Thus, the injection of a solute into a flowing liquid in an empty tube will result in a broadened peak, the width of which can be used to calculate the diffusion coefficient of the solute in the liquid. It follows that experimental measurements would involve only *accurate* determinations of the plate height and the mobile phase average velocity. It should be noted that all general procedures for the operation of chromatographs, outlined in Chapters 3 and 5, are essential for the accurate H determinations, and the effects of temperature, delivery systems, injection procedures on accuracy and precision ought to be also taken into account. In this technique the chromatographic column is an empty tube, therefore, the velocity of the mobile phase is obtained by dividing the tubing length by the mean residence time of the solute in the column. The plate height is obtained from the chromatogram, using the expression

$$H = l w_B^2 / 16 t_r^2 \tag{4}$$

where w_B is the baseline width of the peak in sec
and t_r is the residence time of the solute in the column in sec.

For better accuracy, H can be calculated with the aid of an appropriate interface and a computer. In this case it is best to obtain the first moment as well as the second central moment, which are defined respectively as

$$M_1 = \Sigma x_i y_i \Delta x / \Sigma y_i \Delta x \tag{5}$$

$$M_2 = \Sigma (x_i - M_1)^2 y_i \Delta x / \Sigma y_i \Delta x \tag{6}$$

where x_i is the ith sampling (digitizing) period,
y_i is the peak height at that period,
Δx is the sampling interval
and M_1 is the first moment (center of gravity) of the peak.

The second moment (M_2) thus obtained is the temporal based moment, and the plate height is then calculated as:

$$H = lM_2^2/M_1^2 \tag{7}$$

Requirements for the Experimental System

Giddings and Seager (17) were among the first to realize that gas chromatographic instrumentation can be utilized for such measurements. They named this method the *Chromatographic Broadening Technique* and used it to obtain gaseous diffusion coefficients of several binary mixtures of gases (17-19). Other groups have also used the method (20-23), and at least two reviews (24,25) were devoted to the gas chromatographic version of the technique.

The extension of the method to liquid systems via liquid chromatography is rather obvious. It is surprising, therefore, that the transition was fairly slow to come. Ouano (26), Pratt and Wakeham (27) and Grushka and Kikta (28) were among the first to point out the feasibility of using liquid chromatographic equipment for the measurement of diffusion coefficients in liquid mixtures.

The diffusion coefficients of liquids are about four orders of magnitude smaller than those of gases, therefore, special attention must be given to the experimental setup and certain precautions must be taken if *accurate* and *precise* results are to be obtained. Basically, the requirements of the system are as follows: firstly, the column should be sufficiently long in order to minimize end effects, and to ensure adequate sampling by the solutes of all cross-sectional stream lines of the mobile phase. Secondly, the injection should be made as an infinitely narrow plug, and directly onto the column, thirdly, the detector should be part of the column itself, and finally, the liquid flow in the tubing should be laminar in nature. In practice, however, often only the first requirement, that of sufficiently long column, is met since the rest of the requirements is difficult to fulfill. Frequently, the injection is made either with an injection valve which is connected via a narrow tube and connectors (unions) to the column and the actual injection occupies a finite volume. The detector is not part of the column; rather, it is connected to it with narrow tubing and appropriate connectors and the detector cell has a finite volume associated with it. These factors can contribute significantly to the overall width of the solute zone. These contributions, which will be termed here *extra-column broadening effects*, will adversely influence the accuracy of the measured diffusion coefficients. They must, therefore, be understood and corrected for either experimentally or computationally. For this purpose a detailed description of each possible extra-column effect will now be given.

Sources of Extra-Column Broadening Effects

The introduction of the sample is most often done via an injection valve. The solute enters the column as a plug, usually having the volume of the sample loop in the valve. The contribution of this *finite sample size* to the overall peak variance is given by the following expression (29):

$$\sigma^2 = V_i^2/12Q^2 \qquad (8)$$

where σ_I^2 is the variance due to the finite injection volume in time units,
V_i is the injected volume
and Q is the volumetric flow rate.

Table 1 shows typical values of this contribution for several injection volumes and flow rates. Equation (8) also permits the maximum allowable injection volume for a given upper limit error (δ) in the determination of the diffusion coefficient to be estimated:

$$V_{i(max)} = (12\delta)^{1/2} V_D / n^{1/2} \quad (9)$$

where n is the plate number generated by the tube and V_D is the tube volume.

TABLE 1

The Effect of Injection Volume (V_i) and Flow Rate (Q) on the Variance Contribution of the Injector (σ^2_i)

	Q (mL/min)				
V_i μL	0.010	0.020	0.050	0.500	1.0
1.0	8.3E-4	2.1E-4	3.3E-5	3.3E-7	8.3E-8
2.0	3.3E-3	8.3E-4	1.3E-4	1.3E-6	3.3E-7
5.0	2.1E-2	5.2E-3	8.3E-4	8.3E-6	2.1E-6
10.0	8.3E-2	2.1E-2	3.3E-3	3.3E-5	8.3E-6
20.0	0.33	8.3E-2	1.3E-2	1.3E-4	3.3E-5

The values in the table are the variances (in min^2) for the given flow rate and the injection volume.

For example, if the maximum allowable error due to the injection is 5%, then for a typical system, where the tube is 25 m long, having an internal diameter of 0.0125 cm, and the velocity of the mobile phase is 0.5 cm/sec, the injection volume should not exceed 10 μl. If the diameter of the tube is doubled, the injection volume is less critical and a volume up to 90 μl could be introduced into the column before the measured diffusion coefficients are adversely affected. An expression similar to equation (9) was developed by Guiochon (30).

The contribution of the *connecting tubes* can best be dealt with by using the same diffusion equations as those used for the actual measurements.

In general, dispersion arising in the connecting tube could be insignificant since such tubings are often quite short, and much narrower than the diffusion tube. However, the changes in the diameters between the diffusion column and the connecting tubes can result in the presence of stagnant pockets which can contribute to the overall broadening of the solute zone. Sternberg (29) pointed out that such pockets can broaden the zone via two different pathways which he termed *mixing chamber* and *diffusion chamber*. The variance contribution of the mixing chamber is essentially similar to that of the injection process discussed above, thus, the same limitations concerning the volume of the mixing chamber can be applied. The diffusion chamber, on the other hand, behaves in a different manner. Its contribution to the zone broadening can be obtained from the Stoke-Einstein relationship. Assuming again that the maximum allowable contribution of this broadening mechanism is 5%, the maximum volume of the diffusion chamber can be estimated as:

$$V_{D(max)} = 0.05uV_D l/24D_m \quad (10)$$

where l is the length of the connector
and V_D is the diffusion tube volume.

It is seen that due to the small value of the diffusion coefficient in liquids, this mechanism by itself should not have a detrimental effect on the measurement of D_m.

The detector can also contribute to the zone width and thus affect the accuracy of the measurements. In general, the detector contribution can be assumed to be functionally similar to that of a mixing chamber. Typically then, the volume of the detector should be less than 10 µl for a column of 0.0125 cm. i.d. and 25 m. long.

Since the diffusion column is usually quite

long, a typical experimental setup uses a coiled column. This can introduce an additional source of zone broadening due to the so-called "race track effect", where the molecules at outer tracks move faster than the molecules at the inner tracks. Giddings (31) showed that this effect is insignificant provided that the ratio of the coil radius to the column radius is greater than about 10.

Additional Sources of Error in the Measurements of the Diffusion Coefficient

Coiling the diffusion tube may introduce an effect which is quite opposite to the "race track effect" in that it can narrow the solute zone which, in turn, can lead to a significant error in the determination of the diffusion coefficients. This effect is called *secondary flow phenomenon*. At high flow rates in coiled tubes, there is a transfer of momentum between the liquid and the tube wall which results in a flow component in the lateral direction. This transverse flow component narrows the solute zone, thus introducing an error in the calculation of the experimental diffusion coefficients. Andersson and Berglin (32) as well as Atwood and Goldstein (33) discussed the effect of secondary flow on the measured values of the diffusion coefficients. They found that the diffusion coefficient was independent of the flow rate up to a certain limit, above which the values of D_m experimentally measured increase with a further increase in the flow rate. From these two works and the work of Wakeham (34) the upper limit of the flow rate can be estimated as:

$$Q < 100\pi^2 D_m \eta r^5_{(coil)}/r_t^2 \quad (11)$$

where $r_{(coil)}$ is the coil radius,
η is the viscosity of the liquid
and the other symbols have the meaning previously ascribed to them.

Below that limit, the effect of the secondary flow due to the coiling of the diffusion tube should be minimal.

Other sources of possible errors can be traced to the *geometric limitation* in the diffusion tube itself. By this we mean either variations, systematic or otherwise, in the tube diameter or in its shape. Both these contributions were dealt with by Wakeman (34) and Atwood and Goldstein (33). Both groups found that diffusion values obtained from noncircular tubing are very close to the expected values, provided that the deviation from circularity is not too great. The effect of variation in the cross section of the tube is also negligible if that nonuniformity is not too large. For example, Atwood and Goldstein (33) showed that if the cross-sectional variation over the length of the tube is 20%, the observed diffusion coefficient would be only 0.33% lower than the true value. Thus, with high quality tubing this effect should be insignificant.

Fluctuations in the flow rate of the mobile phase can also be a source of error in the determination. The measurement of the diffusion coefficient depends on the flow rate in several ways. Firstly, the detection of the solute zone occurs only at the end of the column and during a brief length of time. Fluctuations in the flow during this time will result in a peak width which is not a true representation of the average flow rate during the whole experiment. Secondly, the dispersion process occurring in the column is a strong function of the flow rate according to equation (1) that was derived assuming that the mobile phase velocity is constant throughout the experiment. Thirdly, the diffusion coefficient is a function of pressure that may vary with the flow rate. However, modern LC pumps that control the flow in a chromatographic system are capable of delivering the flow not only precisely but accurately, provided that necessary precautions are taken (Chapter 3). It follows that the errors associated with fluctuations in the flow rate of the mobile phase can be rendered insignificant.

Finally, *the effect of the length* of the actual diffusion tube must also be considered. It was mentioned previously that the diffusion tube should be of sufficient length to allow minimization of ends effects and ample sampling of the various cross-sectional streamlines of the mobile phase. Andersson and Berglin (32) as well as Atwood and Golay (35) examined this requirement and its ramifications. They showed that the minimum column length must produce a certain minimum number of plates (n) before the values of the diffusion coefficients can be trusted. This number varies according to the treatment from about 15 (34) to 30 (35). The minimum column length (l_{min}) is given then by the expression:

$$l_{min} = n r_t^2 u / 24 D_m \qquad (12)$$

where all symbols have the meaning previously ascribed to them.

As an estimate of the minimum column length, we shall consider a column having a radius of 0.0125 cm operated at a velocity of 0.5 cm/sec and producing 30 theoretical plates. Under such conditions, l_{min} is equal to about 10 cm. This calculation, however, does not take into account the effects of the extra-column contributions which in essence put further demands on the minimum allowable column length. Nonetheless, the calculation shows that the present practice of using very long columns may be counter productive, and that the measurement time could be further decreased.

All sources of errors discussed above can be minimized, and with the proper instrumental system, the chromatographic broadening technique can yield *accurate diffusion data*. Grushka and Kikta (28), and

Atwood and Goldstein (33) compared some chromatographically obtained results with other methods of measurements. Their results show that the agreement is very good. Moreover, since the dynamic chromatographic method can provide *absolute* measurements, the results indicate its *superiority* as an experimental method.

Recent Applications of the Method

The chromatographic method of the determination of the diffusion coefficients is, albeit slow, gaining acceptance, and applications to many areas of measurement can be found in the literature. We shall present here a short review of selected applications which demonstrate the versatility and reliability of the technique.

Four papers that deal with fundamental aspects of the methodology (32, 33, 34,36) have already been discussed. The work of Atwood and Goldstein (33), however, should be mentioned again since it demonstrates that with careful experimental design very quick measurements can be made. They reported typical measurement times of the order of 15 min. and a rate of data accumulation of 10 triplicates per hour. Claessens and van den Berg (37) reported the use of a straight 24.75 m. long tube, in which the measurement time could be reduced to three minutes. While these workers did not provide sufficient information to allow the accuracy of the results to be determined, the paper nonetheless demonstrates the tremendous potential of the method.

In a recent paper by Katz and Scott (38), the D_m values for about 70 solutes in one solvent system have been reported. The measuring time was only of the order of 2 min. per one diffusion point. Furthermore, the D_m values obtained were utilized to establish the relationship between the solute diffusion coefficient and its molecular weight.

Dymond (39) used the chromatographic broadening technique to obtain the mutual diffusion coefficients of benzene, toluene, p-xylene, mesitylene, naphthalene, anthracene and rubrene in n-hexane. His results agreed quite well with the values of the diffusion coefficients obtained by others and by different measuring techniques. The data obtained was utilized to compare with predictions of the rough hard-sphere (RHS) theory. The results show that while the organic molecules are not spherical, the RHS model can yield satisfactory predictions of the diffusion coefficients, provided that the solvent's translation-rotation coupling factor is used. Typical measuring times in this work were of the order of 90 min. per one diffusion point.

Chan (40) also examined the use of the RHS model for the interpretation of diffusion data. He used the chromatographic broadening technique to measure the diffusion coefficients of the pseudospherical molecules of CCl_4, CH_3CCl_3, $(CH_3)_2CCl_2$ in acetone, methanol, ethanol and n-decane. It was found that in general the D_m values for each solvent were independent of the molecular mass or dipole moment of the solutes. This was taken as a further justification of the basic assumptions leading to the RHS model. In a second paper Chan (41) elaborates further on this topic. In this paper the diffusion coefficients of o-,m-, and p-dichlorobenzene, o-,m-and p-xylene and chlorotoluene in acetone, ethanol and n-tetradecane were measured. The trend of the results was similar to that found in his previous work: the diffusion coefficients were insensitive to the molecular mass and dipole moment. A dependency on the geometry of the solutes was found, and this dependency was a function of the solvent used. While he pointed out that the RHS model can be employed to explain the results, no attempts were made to present such an explanation. The rate of data collection in these two papers was between 6 and 12 hours per one diffusion coefficient.

Tominaga and his co-workers used the chromatographic broadening technique to measure the diffusion coefficients of benzene, toluene, ethylbenzene and hexafluorobenzene in water (42) and of benzene, n-butylbenzene, biphenyl, naphthalene and 1-ethylnaphthalene in water (43). In the first of the two papers, the measurements were made in the temperature range from 298K° to 368K°, while in the second work the range was from 265K° to 433K°. The papers demonstrate nicely one advantage of the chromatographic technique: it overcomes the difficulties, encountered in conventional methods, in measuring the diffusion of non-polar, slightly soluble solutes in water. Tominaga and his co-workers found that the activation energy for the diffusion decreased with increase in temperature. At high temperatures, the hard sphere theory could be used to explain the behavior of the diffusion data. However, discrepancies were found at low temperatures. They also found that the Stokes-Einstein relation holds much better in water than in organic solvents. They attributed this behavior to counterbalancing effects arising from the enhanced water structure. However, they were unable to detect experimentally any positive evidence for the effect of the enhanced water structure on the diffusion coefficients. The rate of data collection in these experiments was between 30 to 45 min. per one D_m value.

An interesting extension of the chromatographic broadening technique was carried out by Walters and co-workers (44). They used the method to determine the diffusion coefficients of proteins in aqueous solvents. The agreement between the D_m values for 16 different proteins and those reported in the literature was very good, with typical deviations being less than 10%. From the diffusion data the authors were able to obtain a good correlation with the molecular weights of the proteins. The results clearly indicate the potential of the method for measuring diffusion coefficients of large (polymeric) molecules. Walters et al. used a 2.8 m. long tube having a radius of 0.025 cm. for the diffusion measurements. Typical measurement times were 30 to 90 min.

All recent applications of the technique have thus demonstrated that chromatographic equipment can be used to accurately measure diffusion coefficients in gases and in liquids. Furthermore, with modern LC units, which can include data acquisition, reduction and treatment by a computer, diffusion coefficients in liquid mixtures can be obtained in a matter of minutes. Provided that the experimental setup is properly designed, no other method can rival the chromatographic broadening technique in the ease of data collection, and in the *accuracy* and *precision* of the *determined diffusion coefficients.*

ADSORPTION ISOTHERMS

Physico-chemical characterization of solids has always been an active area of research done in many fields of chemistry. Advantages of dynamic chromatographic methods (e.g., precision and speed) over conventional static techniques have been extensively utilized for surface characterization, for example, by *adsorption isotherms.* There are quite a few papers dealing with this topic in connection with gas chromatography. The monograph of Conder and Young (2) treats this subject rather extensively. Liquid chromatography, on the other hand, has not been often used to investigate the adsorption phenomenon. The remainder of this chapter will attempt to fill the gap. For this purpose we shall include a brief review of the theory and then describe a few papers dealing with the subject.

When two immiscible phases, containing several components, are brought into contact, the composition of one or both phases may change due to redistribution of one or more of the components. This phenomenon occurs if both phases are liquids or if one phase is solid and the second is liquid or gas. In the case when one of the phases is a solid, the process of accumulation of molecules on the interfacial layer is called adsorption. One way to quantitate the relationship between the concentration on the surface and in solution is via the adsorption isotherm. The ratio of the concentration of the species in the two phases is defined as the distribution coefficient, and the isotherm describes the variation in that distribution coefficient as a function of the solute concentration. In this manner, the adsorption isotherm can serve as a *thermodynamic descriptor* of the *distribution equlibria* involved in the chromatographic processes. Therefore, the quantitative description of the adsorption isotherm is essential for a better understanding of the nature of the various interactions in the chromatographic column between the solute and the mobile and stationary phases, as well as between the phases themselves.

In general, it can be stated that the concentration of solute (c_s) adsorbed on the stationary phase is a function of that in the mobile phase (c_m) and of the temperature. At a constant temperature, this general form of the isotherm can be written as:

$$c_s = f(c_m) \qquad (13)$$

Typically, the experimental approach is to determine this function at a given temperature. Ideally, the isotherm should have the following linear form:

$$c_s = Kc_m \qquad (14)$$

where K is the distribution coefficient.

K is related to the capacity factor (k') via the phase ratio (θ):

$$k' = K\theta \tag{15}$$

Figure 2 schematically shows a linear isotherm.

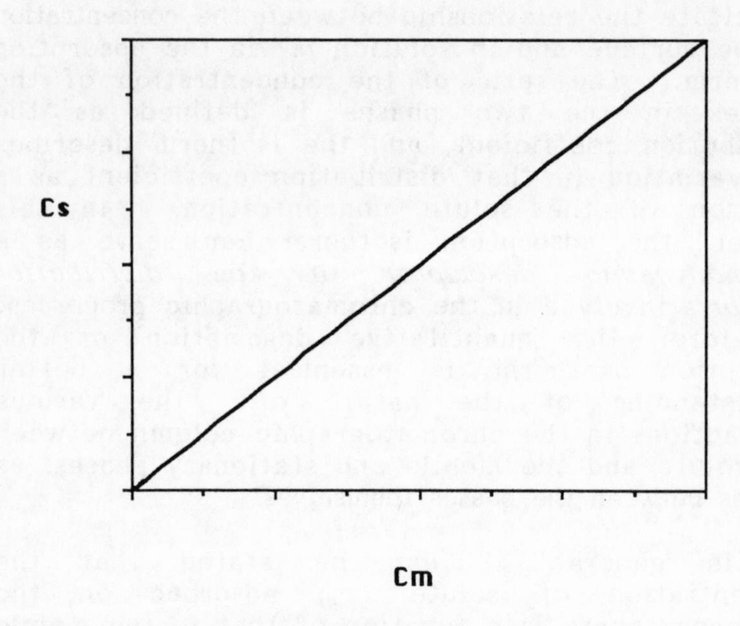

Figure 2. Linear isotherm.

Frequently, however, the isotherm is not linear. Various functional expressions of the adsorption isotherm have been proposed, either as a result of some empirical observations or in terms of a specific model. Perhaps the best known expressions are those developed by Langmuir and Freundlich . The Langmuir model (45) assumes a constant heat of adsorption, a finite number of active sites on the surface where adsorption takes place, and it also assumes that the solute molecules interact only with the adsorption sites and not with one another. The meaning of these assumptions is that the maximum adsorption corresponds to a saturated monolayer of solute molecules on the adsorbent surface. The Langmuir isotherm can be written as:

$$c_s = ac_m/(1+bc_m) \qquad (16)$$

where b is a constant related to the energy of adsorption

and a is the slope of the isotherm at low solute concentrations and it is equivalent to the distribution coefficient (K).

A typical Langmuir isotherm is depicted in Figure 3.

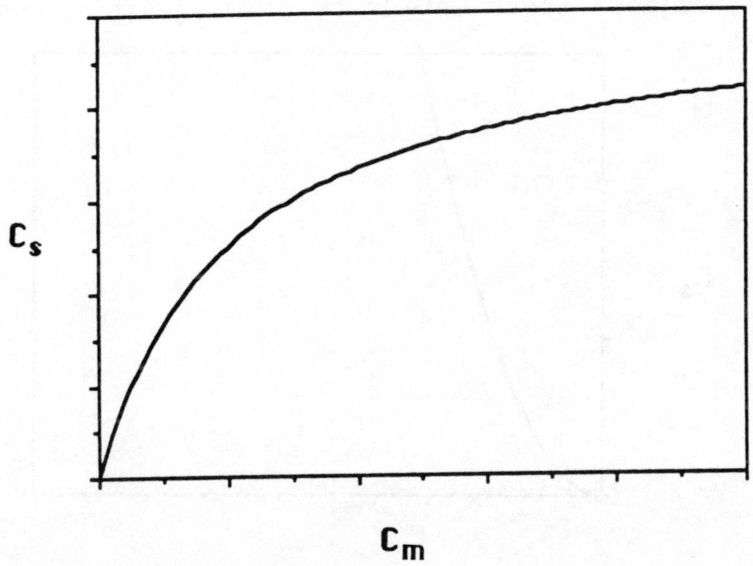

Figure 3. Langmuir isotherm.

The Langmuir equation, which was derived theoretically (45), has at least two attractive features often observed experimentally. At low concentrations of the solute, the equation predicts a linear relationship between the amount of solute adsorbed to that in solution, from which a value of the distribution coefficient can be obtained. At the high solute concentrations, the equation predicts an asymptote (a/b). Consequently, this useful physico-chemical constant, the *limiting monolayer concentration*, can also be determined.

The Freundlich approach (46) is more empirical in nature, and it can be written as:

$$c_s = a(c_m)^n \tag{17}$$

where a is related to the adsorption capacity,
and n is the "intensity" of the adsorption.

Both these parameters are empirical, and they are determined experimentally by plotting log c_s versus log c_m.

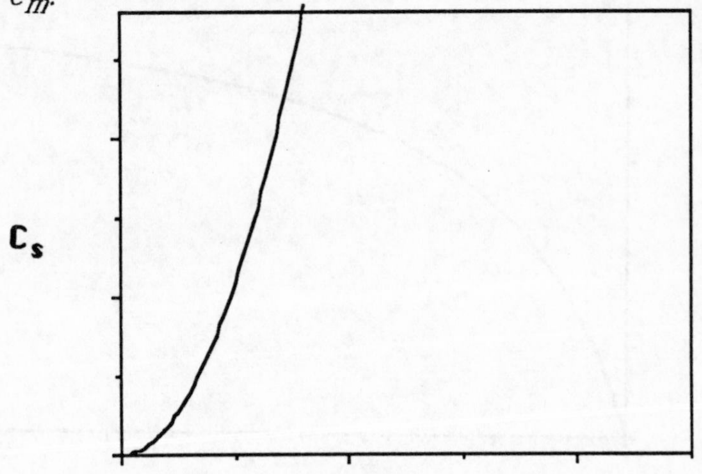

Figure 4. Freundlich isotherm.

Unlike the Langmuir isotherm, the Freundlich equation does not predict linear behavior at low solute concentrations, nor the asymptotic behavior at high solute concentrations. Figure 4 shows a typical Freundlich isotherm.

In addition to the isotherms shown in Figure 2-4 there are several other forms, the most important of which is demonstrated in Figure 5.

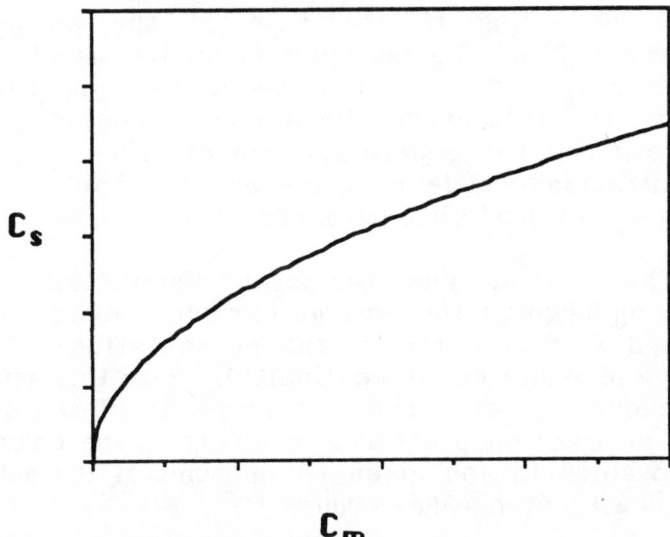

Figure 5. Modified Langmuir isotherm.

This isotherm, which is often found experimentally, can be described by suitably modifying the Langmuir and Freundlich expressions, as well as from other equations such as power expansion of c_S (47,48). The isotherm shown in Figure 5 is important especially in connection with overloaded conditions.

The Chromatographic Peak Shape and Its Dependence on the Adsorption Isotherm

The effect of the *peak shape* on resolution and data processing and consequently on the accuracy and precision of experimental data has already been discussed in Chapters 3 and 2. In this chapter the importance of the peak shape will be again considered since it is a reflection of the thermodynamic and kinetic factors involved in the chromatographic process (49). The solute is injected onto the column as a narrow zone. During the chromatographic run, this zone broadens and acquires a specific shape. The front or rear boundary of that zone can be *sharp* or *diffuse* according to the type of the adsorption isotherm. Thus, the isotherm is an important factor in determining the overall peak shape. In order to examine the relationship between the peak shape and the isotherm, the equation which describes the solute concentration profile in space and in time (i.e. the elution curve) must first be discussed.

The spatial and temporal distribution of a solute undergoing the chromatographic process can be obtained from the law of the conservation of mass (using the equation of continuity). For this purpose an accounting of various solute-solvent-stationary phase interaction processes occuring in the column is needed since in any given volume unit of the column, the following events may occur:

1) solute molecules enter the column volume equilibrated with the mobile phase;
2) some of the solute molecules will be retained by the stationary phase;
3) some of the solute molecules may reenter the mobile phase from the stationary phase;
4) solute molecules are eluted from the column volume.

One approach to the understanding of the distribution process of the solute in the column is carried out in the following manner. Consider the volume segment of the column between two cross sections, one at x and the other at $x+dx$. The total concentration (c) of the solute in this volume segment is the sum of the concentrations of the solute in the mobile and stationary phases. The ratio of the two concentrations, of course, is determined by the distribution coefficient. The amount of solute which is allowed to be in the mobile phase is $R_m c$. R_m is then the fraction of the solute molecules in the mobile phase. During the time interval (dt), the volume of the mobile phase entering and leaving this volume is equal to udt, where u is the mobile phase velocity. During this time, the net gain of material in the volume is the difference between the amount entering and leaving, that is:

$$\text{net gain} = R_m udtc_x - R_m udtc_{x+dx} \tag{18}$$

Subscripts x and $x+dx$ indicate quantities at the two cross sections mentioned above. The term c_{x+dx} can be expanded in a Taylor series relative to x to give:

$$c_{x+dx} = c_x + (dc/dx)_x dx + \ldots \tag{19}$$

Expansion terms higher than the second were neglected, thus, the net gain in the solute is:

$$\text{net gain} = -R_m udt(dc/dx)_x dx \tag{20}$$

Equation (20) can be rewritten as:

$$(\text{net gain})/dtdx = -R_m u(dc/dx)_x \tag{21}$$

The left-hand side of equation (21) is essentially dc/dt since dx is the volume element between the two cross sections, and, therefore, the ratio of the amount gained to dx is the change in the concentration. Moreover, from the definition of the material flux (J),

$$J = uc \tag{22}$$

it is seen that the right-side of the equation is the distance gradient of the flux. Thus, equation (21) reduces to:

$$dc/dt = -R_m(dJ/dx) \tag{23}$$

When dispersion of the solute is significant, the flux is also equal to:

$$J = uc - D_p(dc/dx) \qquad (24)$$

where D_p is the dispersion coefficient.

The combination of equations (21) and (22) yields the sought after expression which describes the behavior of the solute zone as a function of time and distance:

$$dc/dt = -R_m u(dc/dx) + R_m D_p(d^2c/dx^2) \qquad (25)$$

Since $R_m c$ is the fraction of the solute in the mobile phase, this equation can be written as:

$$dc/dt = -u(dc_m/dx) + D_p(d^2c_m/dx^2) \qquad (26)$$

This second order differential equation, together with the appropriate initial and boundary conditions, allows the concentration profile of the solute eluting from the column to be obtained. When the distribution isotherm is linear; that is, dc_s/dc_m is the constant K, the solution of the last equation is in the form of a Gaussian peak. That, however, is not the case when the isotherm is not linear; i.e, dc_s/dc_m is no longer a constant. The reasons for the "nongaussian" nature of the peaks hinge on the fact that the transformation between c and c_m in equation (25) is no longer straightforward due to the nonlinearity of the isotherm. While in this case the solution of the differential equation is complicated, a feeling of the distortion of the solute zone can be obtained from the following arguments.

The solute zone is moving at a velocity $U = R_m u$, where, as mentioned previously, R_m is the equilibrium fraction of the solute in the mobile phase. That fraction can also be expressed as

$$R_m = 1/(1+\theta dc_s/dc_m) \qquad (27)$$

where θ is the phase ratio
and dc_s/dc_m is the adsorption isotherm slope.

In the case of linear chromatography, the slope is simply the distribution coefficient (K). The importance of equation (27) can be demonstrated in the following manner. Assume that the solute zone in the column is divided into many small segments. The average velocity of each segment is dictated by the relationship shown in equation (27). As long as the isotherm is constant, that is, K is independent of the solute concentration, the velocities of all the segments will be identical, and the eluting peak will be a Guassian. However, when the isotherm is not linear, each segment of the solute zone will move with a different average velocity, thus, resulting in asymmetric peaks. The degree of the asymmetry is, of course, a function of the nature of the nonlinearity of the isotherm.

The Langmuir Type Isotherms

These isotherms show a *saturation* effect of the stationary phase: above a certain concentration of the solute in the mobile phase, c_s changes very slowly or not at all with any further increase in c_m. That means that the slope of the isotherm decreases as the amount of solute is increased. In this case, if the solute segments in the column are examined, the following will be observed. In the front part (closer to the column exit) of the zone, at segments having very small concentrations of the solute, the velocity of each segment will be constant. However, as we examine segments closer and closer to the zone maximum and past a certain local concentration of the solute, the slope of the isotherm is no longer a constant, but rather a decreasing value at each successive segment. Equation (27) indicates that as the slope of the gradient is decreased, the velocity is increased. Thus, the velocity of each segment is faster, the closer it is to the peak maximum. This is demonstrated in Figure 6, in which a solute zone is drawn, and the velocities of all the segments are schematically indicated with horizontal arrows. The magnitude of the velocity vectors is given by the length of the arrows.

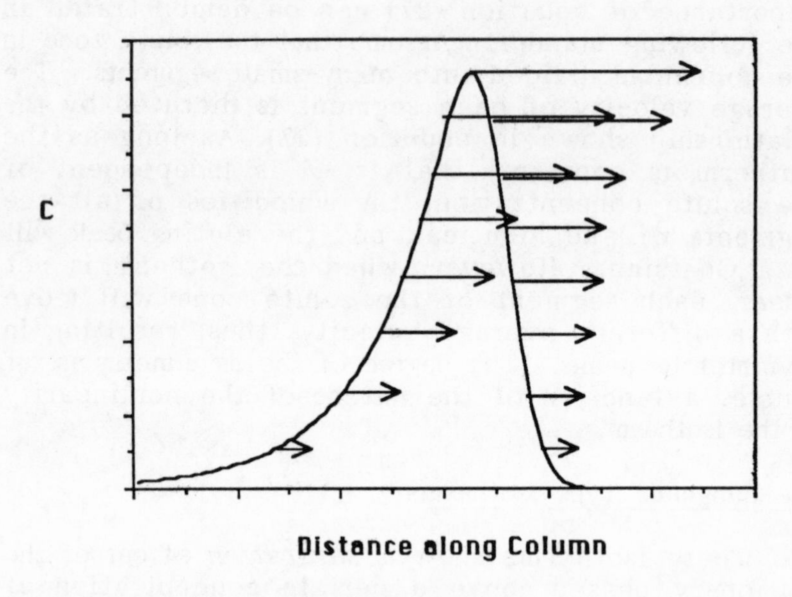

Figure 6. The generation of the peak shape in the case of the Langmuir type isotherm.

The net effect of this behavior is a self-sharpening of the front of the solute zone. The rear part of the zone is broadened asymmetrically since the top part of the back half moves faster than the bottom half. In other words, the rear boundary becomes more diffuse as points of lower concentrations are moving slower than those with higher concentrations. On the chromatogram the peak will show a fast rising front and a slowly decaying tail (Figure 6). It should be mentioned, perhaps, that the chromatogram as seen by the detector is the mirror image of the zone shape in the column. This is due to "distance to time" coordinate transformation involved in the detection process: the detector is stationary at the end of the column, and the solute zone is swept through the detector as a function of time.

The practical aspects of the above discussion are rather important. The implication of Langmuir type isotherms is the presence of strong adsorption. Therefore, the injection of a very small sample can lead to irreversible adsorption and the absence of the peak from the chromatogram. Moreover, as the amount of the injected solute is increased, the peak shape is apt to change, becoming more tailed as the concentration is increased. A corollary of this is a *constant shift* in the peak maximum, making the *measurement* of the retention times *prone* to *errors*.

Freundlich Type Isotherms

In the classical sense these isotherms are *concave* to the c_m axis; that is, as the concentration of the solute is increased, there is a point at which c_m does not change greatly. In other words, the slope of the isotherm increases with increasing amounts of solute in the column. The shape of the solute zone is the reverse of that described above. If the zone is considered to be divided into small segments, the following effects will be observed. On the front part of the zone, at very low concentrations, the slope of the isotherm is small and, therefore, the velocities of these segments are high. As segments nearer the zone maximum are examined, it is observed that their velocities decrease continuously as a consequence of continuously increasing isotherm slope (see equation 27). Hence, this boundary of the zone is self-diffusing. Past the maximum, the velocities of the segments again increase demonstrating self-sharpening. The net effect is that the solute peak in the chromatogram has a slow increasing front and a fast descending tail. Figure 7 shows the zone shape. Here, as with the Langmuir isotherm, the peak shape and the maximum position are a strong function of the amount of the solute injected. This type of behavior is often seen when the chromatographic column is overloaded.

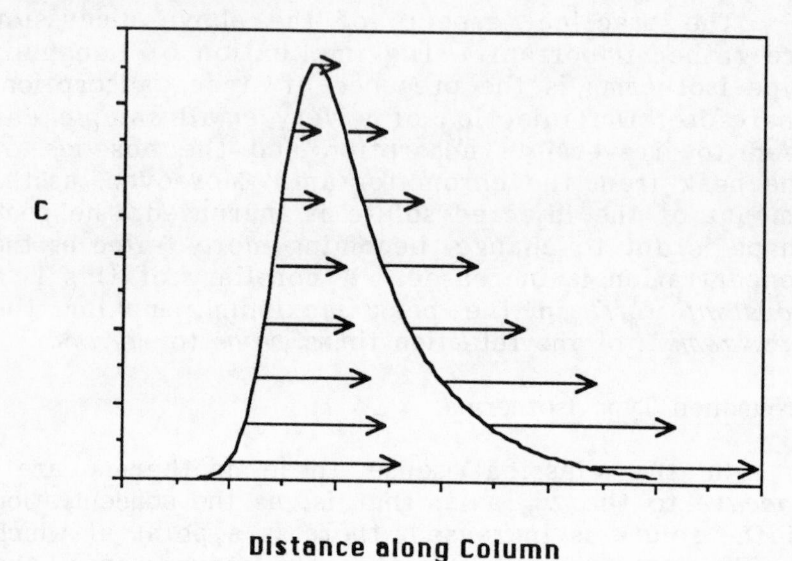

Figure 7. The generation of the peak shape in the case of the Freundlich type isotherm.

An interested reader is referred to the excellent monograph by Hellfrich and Klein (50) for a more comprehensive treatment of the dependence of the peak shape on the isotherm.

The above discussion demonstrates the connection between the adsorption isotherm and the peak shape. The rule of reciprocity works here as well, and it follows that from the peak shape, or any other chromatographic signal, information about the isotherm and its shape can be obtained. It has to be again pointed out that all necessary precautions, mentioned previously and discussed in Chapters 2 and 3 in the operation of the LC system, should be considered if the precise and accurate determination of the adsorption isotherm is to be obtained. However, another point needs to be clarified since it will also affect the *reliability* of the measurements and their interpretation; that is, column void volume and its determination.

Adsorption Isotherm and Void Volume Determination

The determination of the column void volume (V_O) is essential to all thermodynamic measurements accomplished by chromatography. The problem of obtaining an accurate and unequivocal result for this parameter has not yet been solved. Many experimental approaches have been put forward to measure the void volume. These methods include: 1) weighing the column with and without the mobile phase, or with two different solvents (51), 2) linearization of the retention parameters of a homologous family of solutes (52,53), 3) injection of a salt or anion (54), 4) injecting one of the eluent components (55) etc. Melander and co-workers (56) as well as Knox and Kaliszan (57) recently surveyed the various methods for the calculation of the void volume. It is not the aim of this chapter to provide an exhaustive discussion of this crucial issue. In the context of this chapter, however, the void volume is related to the adsorptive capacity of the stationary phase and therefore to the isotherm of the various components in the mobile phase.

The problem in the measurements of the void column lies in the difficulties of determining the exact interface between the mobile and stationary phases. It might be assumed that the column void volume is essentially the geometric interstitial volume. In fact, for a single component mobile phase the two volumes can be identical as the density and partial molar volume of the solvent are not changed in the adsorption layer, if such exists. In multicomponent mobile phases the situation, however, is much more complex. In such cases an adsorption, or solvation, layer is usually present. This layer can have a different composition and a different density than the bulk mobile phase. However, the demarcation between the bulk mobile phase and this solvation layer is a matter of conjecture. Several approaches have been suggested to resolve this question. Kovats and his co-workers (58,59) were the first to attempt a rigorous theoretical study of the issue. They used the approach of Gibbs (60) that utilizes a hypothetical plane which serves as a division between the two phases. Problems still arise concerning the exact location of this plane. Since it is a hypothetical plane, its actual distance from the solid matrix depends on the assumptions made. More recently, Melander et al. (56) have discussed the shortcomings of various conventions used to place the plane. Their recommendation is to employ the retention volume of a labelled isotope of the least retained component of the mobile phase as the void volume marker. Knox and Kaliszan (57) have pointed out the importance of the thermodynamic void volume (V_O) which includes not only the interparticle volume but also that of the pores. Their recommendation is shown in the following equation:

$$V_o = V_A X_A + V_B X_B + \ldots \tag{28}$$

where V_A etc. are the elution volumes of the isotopically labelled components of the mobile phase

and X_A etc. are the volume fractions of the components.

This approach will often yield void volume values different from those advocated by Melander et. al. (56). However, for some mobile phase compositions the differences between the two approaches are minor, and both teams claim that the experimental results of McCormick and Karger (61) support their respective theories.

The work of Knox and Kaliszan (57) is of particular interest in the context of this chapter, since it indicates a direct connection between the retention volumes of isotopically labelled mobile phase components and their partition isotherms between the bulk mobile phase and the space between the particles of the stationary phase. In their paper the authors present the isotherms for several binary mixtures of solvents, some of which are commonly used as mobile phases in HPLC. These isotherms show which of the components is preferentially absorbed on the stationary phase. Table 2 summarizes some of their results. The trends depicted are generally in agreement with those of Snyder's ε^0 values or the Hildebrand's solubility parameters. However, some inconsistencies are apparent, such as the adsorption of ethanol in preference to acetonitrile. Knox and Kaliszan attribute this observation to the effect of the residual silanols.

TABLE 2

Preferential Adsorption of Components in Binary Mixtures

System A	System B	Max. Excess (vol. %)	Composition at Max. Excess
Acetonitrile	H_2O	6.8	0.45
THF	"	6.5	0.35
Ethanol	"	2.6	0.2
CCl_4	Acetonitrile	7.0	0.38
"	Ethanol	2.3	0.5
Ethanol	Acetonitrile	1.2	0.4
Benzene	Ethanol	2.0	0.45

a)Excess values give an indication of the component which is preferentially adsorbed. The larger the value the stronger the adsorption of component A. Taken from ref. 57.

A corollary of the above discussion is that if the adsorption isotherm of a solute is known, then the void volume of the column can be calculated. This is so since the isotherm gives the amount of the adsorbed species on the stationary phase. The following few equations demonstrate the point. The definition of the capacity ratio is:

$$k' = m_s/m_m \tag{29}$$

where m_s and m_m are the weights of the solute in the stationary and mobile phase respectively.

The experimental definition of k' is:

$$k' = (V_R - V_o)/V_o \tag{30}$$

where V_R is the retention volume of the solute. These two equations can be combined to give:

$$V_o = V_R - (m_s/c_m) \tag{31}$$

In equation (31) use was made of the relationship $m_m/V_o = c_m$, where c_m is the concentration of the species in the mobile phase. Therefore, knowing m_s from the adsorption isotherm allows the calculation of the column's void volume. This method of calculating the void volume is, at least theoretically, easy and straightforward. It is attractive since it does not depend on a choice of a suitable marker to be injected onto the column. However, few chromatographers utilize this approach. One exception is the recent paper by Levin and Grushka (62) who describe the use of the isotherm for the calculation of the void volume.

Experimental Measurements of Isotherms

The isotherm, as it has been already shown,

influences the shape of the solute zone while undergoing chromatographic development. In addition, the above discussion indicates that the isotherm can provide useful information directly pertaining to the basic chromatographic process. However, before enumerating further the use of the adsorption isotherms in liquid chromatography, some discussion should be given regarding the actual measurement of the isotherm. The following discussion will concentrate on two major methods for the experimentally measurement of isotherms.

Static Techniques

Conventional measurements of adsorption isotherms were carried out using batch techniques. Examples of this method can be found in many references (63-67). In this approach, a weighed amount of the adsorbent (m^o) is immersed in a solution of known composition. Each component in the solution is characterized by its own mole fraction, (X_1^o). The two phases are brought into an intimate contact by a agitation, until an equilibrium is achieved. Then the composition of the liquid phase is determined by a suitable means, e.g., gas chromatography, and the new mole fractions (X_1) of the various components are recalculated. Any changes from the original composition is due to adsorption. The difference in the mole fractions (X_1^o-X_1) yields one experimental point of the isotherm. The process is then repeated with a new solution having new concentrations of the components.

The main advantages of the batch, or static, technique lies in that the results of the measurements yield the isotherms directly. There are, however, several disadvantages of the static technique:

1) It is time consuming and cumbersome.

2) There is an uncertainty about the time it takes for the system to reach equilibrium.
3) It requires large amounts of adsorbents and solutions in order to insure the reliability of the results.
4) Wetting difficulties cause problems when the attempt is made to measure the isotherms of compounds on reversed phase material.

These difficulties outweigh the few advantages mentioned before, and alternative methods were sought. Some of these are described in the next section.

Dynamic Techniques - Chromatographic Methods

The possibility of using chromatographic equipment to measure adsorption isotherms was recognized early in the days of chromatography (2). It was realized that as with diffusion measurements, if the adsorption isotherm affects the chromatographic behavior of the solutes, then the chromatographic data should yield, under proper conditions, information concerning the isotherm itself. Chromatography, being a dynamic method, offers some clear advantages. Firstly, as a dynamic method, it offers a great saving in analysis time. Moreover, since the mobile phase is forced to perculate through the packing by a pump the problems of wettability of the stationary phases are circumvented. Thus, reversed phase studies in aqueous liquid systems can be accomplished. Another advantage lies in the fact that since chromatography is a method of separation, the solutes do not have to be pure. Liquid chromatography offers several variants of the method.

Frontal Analysis - Breakthrough Curves. A more detailed description of this approach can be found in the literature (68-76). In this technique the column is preflushed with a mobile phase containing a solute of known concentration until an equilibrium is established in the column. Then, a sudden change to a mobile phase containing a different concentration of the solute is introduced to the column. This is time $t=0$ of the experiment. The effluent composition at the column outlet is monitored as a function of the elapsed time until it becomes steady and equals the inlet composition. This is time $t=t_{inf}$ which is the time of the equilibration with the new mobile phase in the column. Figure 8 depicts a typical breakthrough curve obtained in such experiments. The shape of the breakthrough curve is that of a self-sharpening boundary. The net adsorbed quantity of the solute is measured from the area under the curve describing the change in the concentration as a function of time until $t=t_{inf}$. If the initial concentration of the solute in the mobile phase is zero, then the following expression gives the relationship between the amount adsorbed and the other parameters:

$$m_s = (t_{inf} - t_o) Q c_m \qquad (32)$$

where Q is the mobile phase volumetric flow rate.

For ideal chromatography, i.e. without any band-spreading, the front of the boundary maintains a square shape throughout the column. In real cases, however, the shape of the front is that of a sigmoid as shown in Figure 8. The inflection point of the front is taken, in these cases, as a measure of t_{inf}. Equation (32) gives one point on the isotherm. To obtain the complete isotherm, the experiment is repeated with several concentrations of the solute in the mobile phase.

Frontal Analysis by Characteristic Point (FACP). A variation of the above method was described by Glueckauf (77). Here, the procedure is the opposite from the previous approach. A mobile phase containing a high concentration of the solute is equilibrated with the stationary phase in the column.

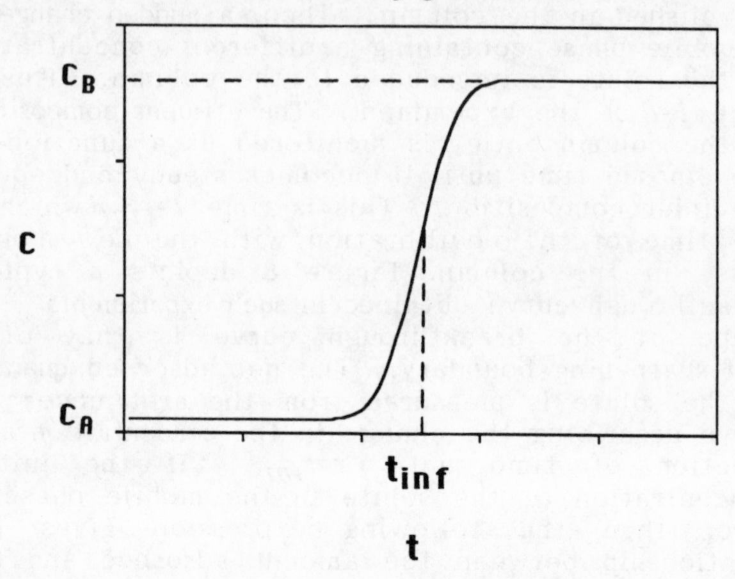

Figure 8. A typical breakthrough curve in frontal analysis.

The mobile phase is then suddenly changed to another mobile phase containing less, or no, solute. The detector monitors the decay of the solute concentration in the column as a function of time. The detector trace is essentially that of a diffuse rear boundary of the solute zone. A typical trace is shown in Figure 9. The main feature of the technique is that it allows the calculation of the whole isotherm from a single experimental run. The amount of solute adsorbed (m_s) on the stationary phase is obtained from the following equation:

$$m_s = 1/V_a \int_{c_B}^{c_A} (V_R - V_o) dc \qquad (33)$$

where V_a is the adsorbent volume.

In the equation the limits of the integration are the two extreme concentrations of the solute. Since the actual solute concentration is needed for the above integration, a detector calibration is required.

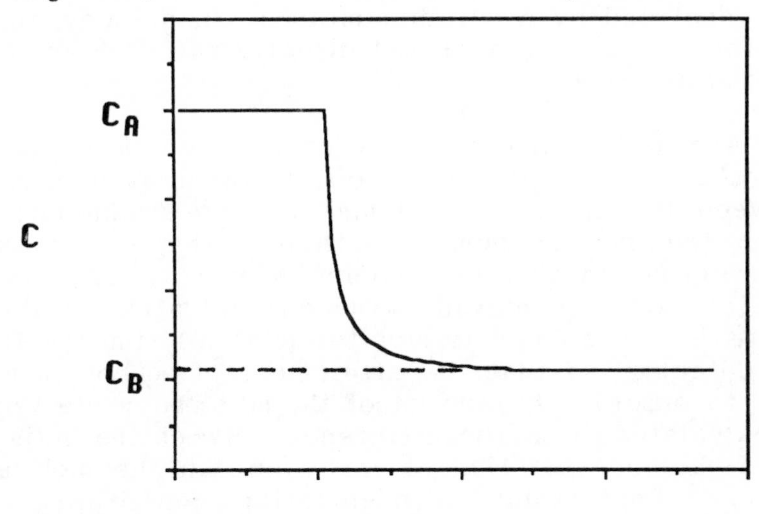

Figure 9. A typical recorder trace of a FACP wave.

This a shortcoming of the method. Another possible source of error lies in that the method does not correct for broadening processes which take place during the boundary elution.

Cremer and Huber (78) introduced a variation of the FACP method, which they called Elution by Characteristic Point (ECP). In this variation, a limited amount of solute is injected so that a elution-like peak results. Equation (31) is then used to calculate the isotherm from the rear boundary of the peak. The main advantage of the ECP method over FACP is that a limited amount of solute is needed for the determination.

Other Methods. Several other procedures for the measurements of the isotherm can be found in the literature. For example, in the *stripping technique* (79-82) the adsorbed solute is displaced from the column by the use of a more retained solvent. The displaced solute is collected into a known volume and then analyzed by a suitable method. In this way, the adsorbed species is analyzed directly rather than by a numerical calculation.

Another method of obtaining the isotherm is by a *recycling technique (83).* This approach is a cross between the static (batch) and dynamic techniques. Here, the mobile phase containing the solute is recirculated through a column several times until equilibration is achieved. The concentration of the solute in the column is monitored at all times with the detector, so that the attainment of equlibrium is easy to observe.The amount of the adsorbed solute can be calculated from the difference between the initial and final concentration of the solute in the mobile phase. Each solute concentration constitutes a single point on the isotherm, therefore, the experiment has to be repeated several times in order to obtain the entire isotherm. The method is quicker than the static approach since the quantitation is done "on the fly" with the chromatographic equipment, and because it is a simple matter to introduce different mobile phases containing varying amounts of the solute. The main disadvantage of the recycling technique lies in the need to calibrate the detector for the quantitative analysis of the solute.

In a recent work, Jacobson et. al (84) examined critically many of the various techniques of measuring distribution isotherms via chromatography. It was their opinion that the method of choice to experimentally obtain the isotherm was by using narrow bore columns run in the frontal analysis mode. Their papers showed the preferred experimental steps needed to ensure the reliability of the results. The precision of their data was very good, so that the temperature dependence of the isotherm could be easily ascertained.

Uses of the Isotherms in Liquid Chromatography

The distribution isotherms can provide useful information as well as an important insight regarding the chromatographic processes which occur in the column. The calculation of the void volume from the isotherm data has been already mentioned. This section will discuss very briefly some other parameters which can be obtained from the isotherm.

Solute Distribution Coefficients

There are several ways by which the distribution ratios of solutes can be obtained with the aid of the isotherms (85, 86). For example, when the void volume is known, then the distribution coefficient can be calculated as follows:

$$K = c_s/c_m = m_s/c_m V_s \tag{34}$$

where V_s is the stationary phase volume increment which results from the adsorption of mobile phase components.

The difficulty with this equation is that V_s is not known. However, V_s can be estimated from the difference between the geometric maximum interstitial void volume (V_o^{max}) and the apparent void volume (V_o^{exp})

$$V_s = V_o^{max} - V_o^{exp} \tag{35}$$

Thus, the distribution coefficient can be calculated from the equation:

$$K = m_s/c_m \ (V_o^{max} - V_o^{exp}) \tag{36}$$

Elucidation of Retention Processes.

Frequently, knowledge of the adsorption of mobile phase components can shed light on the underlying chromatographic processes. In fact, many controversies were solved one way or another with the aid of the isotherms. The following discussion describes some modes of chromatography where the isotherms, either of the mobile phase components or of the solutes, were used to gain a better understanding of the system.

Ion Pair Chromatography. In this mode of chromatography, use is often made of hydrophobic ions in the mobile phase to influence and manipulate the retention of ionized solutes. These large hydrophobic ions are believed to be adsorbed on the alkyl bonded stationary phase or the silica gel surface, forming a primary electrically charged layer. The secondary neutralizing layer is formed by counter ions from the mobile phase.

Two retention processes may be advocated in regard to ionized species. One suggestion is termed "the ion-pair" mechanism, i.e. retention via the formation of an ion-pair between the hydrophobic ion and the oppositely charged solute in the mobile phase (63,87,88), followed by distribution in the stationary phase. The second suggestion is termed "the dynamic-ion-exchange" mechanism (89-93). In this view, the surface of the stationary phase is believed to be coated by the hydrophobic ion-pair reagent. The adsorbed species modify the surface activity so that it is rather similar to a conventional ion-exchange column. The ionized solutes are retained by an ion-exchange mode, displacing the counter ions attached to the functional groups on the surface.

The adsorption isotherms are exploited to evaluate the amount of the adsorbed hydrophobic ion and to elucidate the plausible mechanism of retention. In most cases where reversed phase columns are employed, the indication is that the retention process is dominated by the dynamic-ion-exchange mechanism (72, 89-93). Nevertheless, the adsorption isotherm of the ion-pair reagent by itself is not sufficient for final deduction of the retention process. Thus, the adsorption isotherm of the solutes is also measured simultaneously in these systems (85, 87).

Adsorption chromatography. In this mode of chromatography, there is a formation of an *in situ* stationary phase by the strong adsorption of mobile phase components on the support's surface. This adsorption may change entirely the nature of the original support (66, 94-102). Thus, a normal phase mode of chromatography may be completely reversed by the adsorption of a highly hydrophobic reagent on supports like silica gel (96).

The cardinal question in this mode of chromatography is the nature and construction of the adsorption layer on the stationary phase (99). It could be monolayered (100), bilayered (101), discrete and specific to the various adsorption sites (102), or continuous, more like a surface phase (93,97). Characterization of this adsorption layer may lead to a better understanding of the retention processes in these systems (99). When the adsorption of the mobile phase component is reversible and monolayered, a retention process via the displacement of adsorbed molecules is more probable. However, if the adsorption of this layer is irreversible, or stronger than the solutes, other interactions must be taken into account.

Ligand Exchange Chromatography. This mode of chromatographic separations frequently utilizes metal ions, either adsorbed on the stationary phase (103) or as additives in the mobile phase (103). In such systems the metal ions form complexes with the solutes. The retention of the complexed solutes is different from that of the free solutes. The extent of this change is a measure of the complexing capabilities of the metal ions. In cases where the metal ions are adsorbed on the stationary phase, the amount of the adsorbed metal ions may be correlated to the capacity ratio of the complexed solutes and, thus, the complex stability constant may be evaluated (103).

The distribution isotherms determine the retention behavior of the solutes in any chromatographic system. In addition, however, the isotherm can effect other parameters such as the overall shape of the solute zone in the column. The knowledge of the isotherm is essential in many cases, especially if the operating conditions are to be changed in a desired direction. The best illustration is, perhaps, preparative chromatography, where the operator frequently and deliberately operates in the nonlinear part of the isotherm. In such systems the theoretical calculation of the peak shape and retention becomes a difficult task due to mathematical complexities. Nonetheless, due to the practical importance of nonlinear and nonideal chromatography, a renewed effort is now underway to better understand and characterize such system. In particular, the recent works of Poppe and his group (51,105,106) and Guiochon and his co-workers (107) should be mentioned.

SYNOPSIS

Two important chromatographic parameters, *peak width* and *peak shape*, can be utilized to obtain essential *physico-chemical* properties of liquid-liquid, liquid-solid, gas-solid and gas-liquid systems. The Golay equation describing the solute dispersion in the laminar flow through an empty straight tube can be employed to obtain the *diffusion coefficient values* of gases or liquids. Gas or liquid chromatography equipment is utilized in the *chromatographic broadening technique* to experimentally measure solute diffusivity values. To ensure the *precision* and *accuracy* of experimental results, the chromatographic system is to be carefully designed to meet the following requirements: 1) *instrument dispersion* should be *minimized*; 2) the effect of *secondary flow* ought to be considered; 3) *high quality* tubing of a *circular diameter* is to be used to circumvent the *geometric limitations;* 4) *precise flow-delivery systems* should be employed to minimize flow *fluctuations*; 5) the tube should be *sufficiently long* to produce the minimum number of theoretical plates. Not only can the chromatographic broadening technique yield *accurate* and *precise* diffusivity values, but it also offers simplicity in operation and tremendous time savings, provided that an appropriate system is employed.

Physico-chemical characterization of solids by studying the *adsorption phenomenon* can be carried out utilizing *chromatographic* methods. Adsorption is conveniently described by *adsorption isotherms*, the slope of which is the equilibrium constant of the adsorption process and reflects the nature of interactions between the adsorbed molecules and the adsorbent. The chromatographic *peak shape* is governed by the *type* of the *adsorption isotherm*, therefore, the relationship between the peak shape and the isotherm can be established. There are two major types of the adsorption isotherms, commonly employed, the *Langmuir* type and *Freundlich* type. The Langmuir isotherm, theoretically derived, allows values of *the solute distribution coefficient* and the *limiting monolayer concentration* to be experimentally obtained. Although the Freundlich isotherm does not have a theoretical foundation, it can be useful in describing a specific peak shape. The knowledge of the adsorption isotherm in liquid chromatography is extensively utilized for the *efficient* operation of *preparative* columns. Also, it can be used for the estimation of *the column dead volume.* Experimental methods of measurements of the adsorption isotherms include conventional *static* and *dynamic chromatographic* techniques. The advantages of chromatographic methods are *simplicity* and *speed* in obtaining experimental data, and furthermore, they provide more *accurate* results than the static techniques for *low solute concentrations*. Typical dynamic methods of measurements include the *frontal analysis*, the *frontal analysis by characteristic point*, the *stripping* and *recycling* techniques. The adsorption isotherms are utilized not only for the surface characterization,but also for calculations of the *solute distribution ratios* and *elucidation of retention processes*.

REFERENCES

1. R.J. Laub and R.L.Pecsok, "Physicochemical Applications of Gas Chromatography", Wiley, New York, 1978.

2. J.R. Conder and C.L. Young, "Physicochemical Measurements by Gas Chromatography", Wiley, New York, 1979.

3. R. Kobayashi and T. Kragas, J. Chromatogr. Sci., 23 (1985) 11.

4. N.A. Katsanos and G. Karaiskakis, Advan. Chromatogr., 24 (1984) 125.

5. D.L. Sloan, Advan. Chromatogr., 23 (1984) 91.

6. J.F. Parcher, M.L. Bell and P.J. Lin, Advan. Chromatogr., 24 (1984) 227.

7. D.C. Locke, Advan. Chromatogr., 14 (1976) 87.

8. A.J.B. Cruickshank, B.W. Gainey, C.P. Hicks, T.M. Letcher, R.W. Moody and C.L. Young, Trans. Faraday Soc., 65 (1969) 1014.

9. A.S. Cohen and E. Grushka, J. Chromatogr., in press.

10. S.T. Sie, J.P.A. Bleumer and G.W.A. Rijnders, Sepn. Sci., 1 (1966) 41.

11. M. Suzuki, and J.M. Smith, Advan. Chromatogr., 13 (1975) 213.

12. H.A. Scoble, M. Zakaria, P.R. Brown, and H.F. Martin, Computers and Biomed. Res., 16 (1983) 300.

13. H.A. Scoble, J.L. Fasching and P.R. Brown, Anal. Chim. Acta, 150 (1983) 171.

14. G. Taylor, Proc. R. Soc., London, Ser. A, 219 (1953) 186.

15. R. Aris, ibid, 252 (1959) 538.

16. M.J.E. Golay, in "Gas Chromatography 1958", D.H. Desty (Ed.), Butterworth, London, 1958.

17. J.C. Giddings and S.L. Seager, J. Chem. Phys., 33 (1960) 1579.

18. J.C. Giddings and S.L. Seager, Ind. Eng. Chem. Fundam., 1 (1962) 277.

19. S.L. Seager, L.R. Geertson, and J.C. Giddings, J. Chem. Eng. Data, 8 (1963) 168.

20. J.K. Barr and D.T. Sawyer, Anal. Chem., 36 (1964) 1753.

21. J.F.K Huber and G. van Vought, Ber. Bunsenges. Phys. Chem., 69 (1965) 821.

22. H.J. Arnikar and H.M. Ghule, Int. J. Electronics, 26 (1969) 159.

23. E. Grushka and P. Schnipelsky, J. Phys. Chem., 78 (1974) 1428.

24. E. Grushka and V.R. Maynard, Chem. Tech., 4 (1974) 560.

25. V.R. Maynard and E. Grushka, Advan. Chromatogr., 12 (1975) 99.

26. A.C. Ouano, Ind. Eng. Chem. Fundam., 11 (1972) 268.

27. K.C. Pratt, and W.A. Wakeham, Proc. R. Soc. London, Ser. A, 336 (1974) 393.

28. E. Grushka and E.J. Kikta, Jr., J. Phys. Chem., 78 (1974) 2297.

29. J.C. Sternberg, Advan. Chromatogr., 2 (1966) 205.

30. G. Guiochon, in "High Performance Liquid Chromatography-Vol. 2", Cs. Horvath (Ed.), Academic Press, New York, 1980, p.1.

31. J.C. Giddings, "Dynamics of Chromatography", Marcel Dekker, New York, 1965, p. 46.

32. B. Andersson and T. Berglin, Proc. R. Soc. London, Ser. A., 377 (1981) 251.

33. J.G. Atwood and J. Goldstein, J. Phys. Chem., 88 (1984) 1875.

34. W.A. Wakeham, Faraday Symp., 15 (1981) 145.

35. J.G. Atwood and M.J.E. Golay, J. Chromatogr., 218 (1981) 97.

36. A. Alizadeh, C.A. Niete de Castro and W.A. Wakeham, Int. J. Thermophys., 1 (1980) 243.

37. H.A. Claessens, and J.H.M. van den Berg, J. High Resolut. Chromatogr. Chromatogr. Commun., 5 (1982) 437.

38. E.D. Katz and R.P.W. Scott, J. Chromatogr., 270 (1983) 29.

39. J.H. Dymond, J. Phys. Chem., 85 (1981) 3291.

40. T.C. Chan, J. Chem. Phys., 79 (1983) 3591.

41. T.C. Chan, ibid, 80 (1984) 5862.

42. T. Tominaga, S. Yamamoto and J. Takanaka, J. Chem. Soc. Faraday Trans. 1., 80 (1984) 941.

43. T. Tominaga, S. Matsumoto and T. Ishii, J. Phys. Chem., 90 (1986) 139.

44. R.R. Walters, J.F. Graham, R.M. Moore and D.J. Anderson, Anal. Biochem., 140 (1984) 190.

45. I. Langmuir, J. Am. Chem. Soc., 40 (1918) 1361.

46. H. Freundlich, Colloid and Capillary Chemistry, Methuen, London, 1926.

47. E. Glueckauf, J. Chem. Soc., (1947) 1321.

48. A.W.J. de Jong, J.C. Kraak, H. Poppe and F. Nooitgedacht, J. Chromatogr., 193 (1980) 181.

49. D. de Vault, J. Amer. Chem. Soc., 65 (1943) 532.

50. F. Helfferich and G. Klein, "Multicomponent Chromatography", Marcel Dekker, New York, 1970.

51. E.A. Slaats, J.C. Kraak, W.J.T. Brugman and H. Poppe, J. Chromatogr., 149 (1978) 255.

52. G.E. Berendsen, P.J. Schoenmakers, L. de Galan, G. Vigh, Z. Varga-Puchony and J. Inczedy, J. Liq. Chromatogr., 3 (1980) 1669.

53 A.M. Krstulovic, H. Colin and G. Guiochon, Anal. Chem., 54 (1982) 2438.

54. G.E. Berendsen, and L. de Galan, J. Liq. Chromatogr., 1 (1978) 561.

55. R.P.W. Scott and P. Kucera, J. Chromatogr., 175 (1979) 51.

56. W.R. Melander, J.F. Erard, and Cs. Horvath, J. Chromatogr., 282 (1983) 211.

57. J.H. Knox and R. Kaliszan, J. Chromatogr., 349 (1985) 211.

58. N. Le Ha, J. Ungvaral and E. Kovats, Anal. Chem., 54 (1982) 2410.

59. F. Riedo and E. Kovats, J. Chromatogr., 239 (1982) 1.

60. J.W. Gibbs, Jr., Heterogeneous Equilibria, Trans.Conn. Acad. II, 1878, 382, reprinted in The Scientific Papers of J. Willard Gibbs, Ph.D., LL.D., Vol. I, Dover Publicatins, New York, 1961.

61. R.M. McCormick and B.L. Karger, Anal. Chem., 52 (1980) 2249.

62. E. Grushka and S. Levin, Anal. Chem., 58 (1986) 1602

63. R.P.W. Scott and P. Kucera, J. Chromatogr., 142 (1977) 213.

64. V. Ya. Davydov, A.V. Kiselev and Yu.M. Sapojnikov, Chromatographia, 13 (1980) 745.

65. K.G. Wahlund and B. Elden, J. Liq. Chromatogr., 4 (1981) 309.

66. S. Eksborg and G. Schill, Anal. Chem., 45 (1973) 2092.

67. R. Groh and I. Halasz, J. Chromatogr., 199 (1980) 23.

68. A. Bartha, G. Vigh, H.A.H. Billiet and L. de Galan, J. Chromatogr., 303 (1984) 29.

69. H.L. Wang, J.L. Duda and C.J. Radke, J. Coll. Inter. Sci., 66 (1978) 153.

70. F. Koster and G.H. Findenegg, Chromatographia, 15 (1982) 743.

71. S. May, R.A. Hux and F.F. Cantwell, Anal. Chem., 54 (1982) 1279.

72. A. Bartha and G. Vigh, J. Chromatogr., 260 (1983) 337.

73. J.H. Knox and R.A. Hartwick, J. Chromatogr., 204 (1981) 3.

74. J.H. Knox and J. Jurand, J. Chromatogr., 203 (1981) 85.

75. J.H. Knox and G.R. Laird, J. Chromatogr., 122 (1976) 17.

76. Z. Iskandarani and D.J. Pietrzyk, Anal. Chem., 54 (1982) 1065.

77. E. Glueckauf, Nature, 156 (1945) 748.

78. E. Cremer, and H. Huber, Angew. Chem., 73 (1961) 461.

79. C.R. Yonker, T.A. Zwier and M.F. Burke, J. Chromatogr., 241 (1982) 269.

80. C.T. Hung and R.B. Taylor, J. Chromatogr., 209 (1981) 175.

81. C.T. Hung and R.B. Taylor, J. Chromatogr., 202 (1980) 333.

82. D. Westerlund and A. Theodorsen, J. Chromatogr., 144 (1977) 27.

83. J.L.M. van de Venne, J.H.L.M. Hendrikx and R.S. Deelder, J. Chromatogr., 167 (1978) 1.

84. J. Jacobson, J. Frenz and Cs. Horvath, J. Chromatogr., 316 (1984) 53.

85. K. G. Wahlund and I. Beijersten, J. Chromatogr., 149 (1978) 313.

86. J.M Huen, R.W. Frei, W. Santi and J.P. Thevenin, J. Chromatogr., 149 (1978) 359.

87. A. Tilly-Melin, Y. Askemark, K.G. Wahlund and G. Schill, Anal. Chem., 51 (1979) 976.

88. A. Tilly-Melin, M. Ljungcranz and G. Schill, J. Chromatogr., 185 (1979) 225.

89. R.S. Deelder and J.H.M. van den Berg, J. Chromatogr., 218 (1981) 327.

90. R.S. Deelder, H.A.J. Linssen, A.P. Konijnendijk and J.L.M. van de Venne, J. Chromatogr., 185 (1979) 241.

91. C.P. Terweij-Groen, S. Heemstra and J.C Kraak, J. Chromatogr., 161 (1978) 69.

92. O.A.G.J. van der Houwen, R.H.A. Sorel, A. Hulshoff, J. Teeuwsen and A.W.M. Indemans, J. Chromatogr., 209 (1981) 393.

93. F.F. Cantwell and S. Poun, Anal. Chem., 51 (1979) 623.

94. J.F.K. Huber, M. Pawtowska and P. Markl, Chromatographia, 19 (1984) 19.

95. M. Gazdag, G. Szepesi and M.Hernyes, J. Chromatogr., 316 (1984) 267.

96. Y. Ghaemi, J.H. Knox and R.A. Wall, J. Chromatogr., 209 (1981) 191.

97. R.P.W. Scott and P. Kucera, J. Chromatogr., 112 (1975) 425.

98. Yu.A. Eltekov, Yu.V. Kazakevich, A.V. Kiselev and A.A. Zhuchkov, Chromatographia, 20 (1985) 525.

99. L.R. Snyder and H. Poppe, J. Chromatogr., 184 (1980) 363.

100. R.P.W. Scott and P. Kucera, J. Chromatogr., 149 (1978) 93.

101. R.P.W. Scott and P. Kucera, J. Chromatogr., 171 (1979) 37.

102. L.R. Snyder, "Principles of Adsorption Chromatography", Marcel Dekker, New York, 1968.

103. M. Gimpel and K. Unger, Chromatographia, 16 (1982) 117.

104. B. Feibush, M.J. Cohen and B.L. Karger, J. Chromatogr., 282 (1983) 3.

105. H. Poppe and J.C. Kraak, J. Chromatogr., 255 (1983) 395.

106. E.H. Slaats, W. Markovski, J. Fekete and H. Poppe, J. Chromatogr., 207 (1981) 299.

107. A. Jaulmes, C. Vidal-Madjar, H. Colin and G. Guiochon , J. Phys. Chem., 90 (1986) 207.

LIST OF SYMBOLS

a_m	molar absorptivity
b_c	thickness of the TLC layer
c	solute concentration
c_m	concentration of solute in mobile phase
c_s	concentration of solute in stationary phase
d_p	particle diameter
h	peak height
k'	capacity factor of a solute
$\bar{k}$	mean value of $k'_1 + k'_2 + \ldots$
l	length
m	mass
m_m	mass of solute in mobile phase
m_s	mass of solute in stationary phase
n	efficiency in theoretical plates
n_D	detector noise
r	detector response index
$r_{(coil)}$	coil radius
r_t	tube radius
r_c	column radius
s	slope of linear curve
t_r	retention time
u	mobile phase linear velocity
$u_{(opt)}$	optimum mobile phase linear velocity
v	volume in units of plate volume
w	peak width
w_B	peak width at base
A	multipath factor from the Van Deemter equation
A_p	peak area
B	longitudinal diffusion factor from the Van Deemter equation
C	resistance to mass transfer factor from the Van Deemter equation
C_D	detector concentration sensitivity
D	detector signal
D_m	diffusivity of solute in mobile phase
D_p	dispersion coefficient
F	fluorescence flux

H	height equivalent to a theoretical plate
I	current
I_o	intensity of incident light
I_R	intensity of reflected light
I_T	intensity of transmitted light
J	material flux
K	solute distribution coefficient
K°	temperature °Kelvin
K_A	coefficient of absorption per unit thickness
M_1	first central moment
M_2	second central moment
Q	flow rate
R	detector response
R_c	detector concentration response
R_e	reflectance (signal)
R_F	retardation factor
R_m	fraction of molecules in mobile phase
R_r	electric resistance
R_s	resolution
S	coefficient of light scatter
T	temperature
V_a	volume of adsorbent
V_o	void volume
V_D	tube volume
V_i	sample volume
V_R	retention volume
V_s	volume of stationary phase
W_b	TLC spot diameter
W_w	slit width
X_i	mole fraction
Z_s	migration distance of solute
ε	fraction of column cross section occupied by the mobile phase
η	viscosity
λ	thermal conductivity
σ	standard deviation
σ_c	standard deviation due to column dispersion
σ_e	standard deviation due to extra column dispersion
σ_i	standard deviation due to injection volume

θ	phase ratio
$\emptyset$	quantum yield

Standard Deviation Resulting from Different Sources of Error in TLC Scanning Densitometry

σ_c	due to control of experimental conditions
σ_m	due to measuring process
σ_p	due to spot location
σ_t	due to total error
σ_v	due to sample application

INDEX

Accuracy, precision, HPTLC, 238,251
A/D Converter, 41
Adsorption chromatography, 399
Adsorption isotherms, 360,374
  experimental measurements of, 390
  Freundlich, 378,385
  Langmuir, 377,383
  Langmuir, modified, 379
  linear, 376
  mobile phase components, 389
  uses of, 397
Amino acids, PTH-derivatives of, separation, HPTLC, 206
Amplifier, auto-ranging, 42
Analog-to-digital conversion, 41
Analysis
  active ingredients in drugs, 295
  amino acids, PTH-derivatives, HPTLC, 206
  biological samples, 276-279,300,302,303
  insulins, 297
  metoprolol, HPTLC, 232
  polycyclic aromatic hydrocarbons, HPTLC, 203,208
  precision, LC, 89
Anticircular chromatography, TLC, 196, 210
Application, sample, TLC, 211
Area of peak, measurements, GC, 177-182
Atomic spectroscopic detectors, GC, 142
Automated chromatography systems, need for, 310-312,315,375
Auto-ranging amplifiers, 42
Autosamplers, in chromatography, 311,320
  gas, 326
  liquid, 322,325
  thin-layer, 327
Azobenzene, detector calibration standard, HPTLC, 231,232,234
Band broadening
  empty tube, 361
  packed column, 8,361

Baseline drift, LC, 49
Beer-Lambert Law, 237
Bits, effect on resolution, 41
Calibration procedures, HPTLC, 240,243,245,247
Capacity ratio, 67,376,390
    retardation factor, relationship of, 197
Capillary columns, GC, 172,175,305
Capillary flow, TLC, 198
Cartridges, prepacked
    Bond-Elut™, 285,302
    packing material, 286
    Sep-Pak™, 285,300
Choice of detectors, in chromatography
    gas, 175
    liquid, 70
Chromatograph
    gas, basic system, 20
    liquid, basic system, 21
Chromatographic broadening technique, 364
Chromatographic measurements of adsorption isotherms, 392
Chromatographic separation, principles of, 2
Chromatography
    adsorption, 399
    anticircular, TLC, 196
    circular, TLC, 196
    history, 1
    ion pair, 398
    ligand exchange, 399
    linear, TLC, 196
Circular chromatography, TLC, 196,210
Coarse-particle layers, TLC, 197,199
Column
    capillary, GC, 172,175,305
    efficiency, 14,174,175
    function of, 14
    length, 175
    radius, equation for, 19
    switching, 287,317,318
    void volume, 387,388,390
    void volume marker, 388

Computers in chromatography
archiving, 337
communications, 338
control, 333,334
data acquisition, 43,328
data processing, 330
scanning densitometry, 251
Concentration of sample
biological samples, 286
gas chromatography, 162
water analysis, LC, 76,78
Connecting tubes
effect on extra-column dispersion, 366
Continuous sample development, TLC, 201
Critical pair of solutes, 64,66,173
Data acquisition
automation, 328
frequency, 43,184
Data analysis, automation, 320
Data handling, LC, 38
Data processing, 44, 330
post-processing, 44,50,332
sensitivity, 331
threshold, 45,331
Data storage, archiving, 337
Decomposition of sample
drugs, 274
GC, 160
Degassing procedure, 158
Derivatization of sample
biological samples, 303
gas chromatography, 160
thin-layer chromatography, 254
Detectability of sample, HPTLC, 195,223
absorption mode, 229
fluorescence mode, 229
Detection, automation, 319
Detection methods, HPTLC, 218
absorption, 219,225
flame ionization, 260
fluorescence, 219

image analysis, 260
infrared, 259
laser-based, 260
mass spectrometric, 259
photothermal, 260
Detector
dispersion, 26,367
linearity, 23
role in quantitative analysis, 23
sensitivity, 24
Detectors, GC
alkali flame ionization (nitrogen phosphorus), 130,177,281
argon, 141
atomic spectroscopic, 142
bulk property, 105
calibration, 116
chemiluminescent, 148
choice of, 175
classification, 100
concentration sensitive, 102
contamination of, 176
drift, 111
electrochemical, 108
electrolytic (Hall),138
electron capture, 133,176,303
flame ionization, 127,175,176,281
flame photometric, 137,177
gas density balance, 139
helium, 141
hot wire, 176
infrared, 136
ionization, 141
linear dynamic range, 113
long-term noise, 110
mass sensitive, 102
mass spectrometer, 134
minimum detectable amount, MDL, 103,112,145
multiple, 121
plasma, 141
photoionization, 131,177

reaction-based, 108
response factor, 103,113
short-term noise, 110
signal-to-noise, 111
solute property, 105
spectroscopic, 107
thermal conductivity, 123
thermionic ionization, 140
ultrasonic, 148
ultraviolet absorption, 147
Detectors, LC
choice of, 70
electrical conductivity, 70,72
fluorescence, 55,56,70,278
linearity, 31
multi-channel, 57
multiple, 59
noise, 33,36
performance, 31
refractive index, 71
response, 34
sensitivity, 32
ultraviolet, 55,56,71,277
Development, methods of, 5
displacement, 9
elution, 8
frontal analysis, 5
Diffusion coefficients, liquid, 360,361
function of mobile phase velocity, 362
proteins in aqueous solvents, 373
rough hard-sphere theory, predictions of, 372
Stokes-Einstein relation, 373
Digitization of signal, 41
Diode-array UV detector, 58
Diphenylacetylene, detector
calibration standard, HPTLC, 231,232,234
Dispersion
detector, 26,121
extra-column, 19,365
packed column, 14
Dispersive forces, 4

Displacement development, 9
Distribution coefficient of solute, 359, 376, 377, 382,383,397
Drift, baseline LC, 49
Effect of matrix on
fluorescence response, TLC, 258
method validation, 293
Efficiency,
column, 15, 174-175
measurement of, 15
Electrical conductivity detector, 70,72
Electrochemical detector, 55,56,108
Electron capture detector, 105,119,133,176,303
Elution development, 8,11
Equation
HETP, 15
Kubelka-Munk, 236,237,238,244
mass balance, empty tube, 361
resolution, TLC, 200
Van Deemter, 15,66
Error, effect of
non-Gaussian peaks, 48
peak resolution, LC, 172
sample delivery, GC, 161
signal-to-noise ratio, 48
Error, sources of, in
diffusion coefficient measurements, 370
liquid chromatography, 44
scanning densitometry, 249
Expert systems, 339
External standard method
gas chromatography, 184
liquid chromatography, 87
Extra-column dispersion
definition, 19
detection process, 121
performance of capillary columns, GC, 118
sources of, 365
Fine-particle layers, TLC, 197,199
Flame ionization detector, 127,128,175, 176,260,281
Flow programming, 12

Flow rate, optimum, LC, 68
Flow, secondary, 368
Fluorescence response
 enhancement of, 256
 matrix effects on, 258
 quenching , 255,257,258
Forced-flow, TLC, 194,198
Forces
 dispersive, 4
 ionic, 4
 polar, 4
Fraction collection, automated, 319
Frequency, data acquisition, 43, 184
Freundlich adsorption isotherms, 378,385
Frontal analysis, 5,6
 adsorption isotherm measurements by, 396
Function of the column, 14
Fused peaks, effect on precision, LC, 52
Gas chromatograph, basic system, 20
Gaussian elution curve, 8
Gaussian peaks, 34,45,361
Gradient elution, 12,13
Height of theoretical plate, measurement of, 15
HETP
 curves, 18
 equation, 15
 measurement of, 15
History of chromatography, 1
Incomplete resolution, effect on precision, 50
Infrared spectrometer, 136,259
Internal standard method
 gas chromatography, 186
 liquid chromatography, 86
Introduction of sample
 automation,316
 gas chromatography, 159
 liquid chromatography, 79
 reproducibility, 311
Ion pair chromatography, 398
Ionic forces, 4
Ionization detectors, 130,131,140,141,177

Kubelka-Munk equations, 236,237,238,244
Langmuir adsorption isotherms, 377,383
modified, 379
Layer, solvation, 387
Layers, TLC, 197
coarse-particle, 197,199
fine-particle, 197,199
Ligand exchange chromatography , 399
Limit of detection, 33,145,195,292
Limit of quantitation, 33
Linear adsorption isotherms, 376
Linear chromatography, TLC, 196
Linear dynamic range
absorption mode, HPTLC, 223, 239
fluorescence mode, HPTLC, 245
GC detectors, 113
Linearity, detectors,23,31,72
Liquid chromatograph, basic system, 21
Liquid chromatography, analysis, precision, 89
LOD, limit of detection, 33
Longitudinal diffusion, 16
LOQ, limit of quantitation, 33
Mass balance equation, empty tube, 361
Mass of sample, LC, 84
Mass sensitivity, 25
Mass spectrometer, 134, 259
Matrix effects on
fluorescence quenching, 258
method validation, 293
MDC, minimum detectable concentration, LC, 35
Measurements of
adsorption isotherms, 390
long-term noise, 35, 111
peak area, GC manual, 177
electronic, 183
Mechanism of spot reconcentration, 194,204,205
Methods of quantitation
external, 87,184
internal, 86,186
normalization, 86,88,184
Migration of solvent, TLC, 194

Minimum detectable concentration, MDC, LC, 35
Molar absorptivity of sample, 236,238
Moment, first and second central, 363
Multi-channel detectors, LC, 57
Multipath process, 16
Multiple detectors, 59, 121
Multiple sample development, TLC, 194, 202
Nitrogen phosphorus detector (alkali flame ionization), 130,177,281
Noise
    detection, LC, 33
    effect on retention time, 69
    level, 36
    system, LC, 37
Non-Gaussian peaks, error, 46
Normalization method
    gas chromatography, 184
    liquid chromatography, 86,88
Number of theoretical plates, 198, 209
Operating conditions, LC, 80
Operating procedures, LC, 65
Optimum flow rate, LC, 68
Optimum mobile phase velocity, LC, 66,67
Packed column, dispersion, 14
Peak area, measurements, GC, 177
Peak area/peak height, 39
Peak height, measurements, GC, 182
Peak height/peak area, 39
Peaks
    fused, effect on precision, 52
    Gaussian, 34, 45, 361
    identification of, 54,233,259
    resolution of, 172
    shape of, dependence on adsorption isotherm, 379,384, 385, 386
    standard deviation of, 172, 173
    variance of, 361
Permanent gases, analysis, 162
Photoionization detector, 131
Plate
    efficiency, TLC, 199

material, 193
Plate height, 15,198,361,363
Polar forces, 4
Polycyclic aromatic hydrocarbons,analysis of, TLC 203,204,208
Precision, analysis
drugs, 296
insulins, 298
liquid chromatography, 89
thin-layer chromatography, 238,258
Preparation of sample
automation, 313
drugs, 282
gas chromatography, 157
liquid chromatography, 74
manual, robotic, comparison, 348
reproducibility, 311
robots, 314,341,348
Principles of separation, 2
Programming
flow, 12
gradient, 12
temperature, 12
Proteins, in aqueous solvents, diffusion coefficients of, 373
Quantitative analysis, LC, 63,85
operating procedures, 65
Quantitative analysis, TLC, 236
accuracy and precision, 238
calibration procedures, 238,240,241,245,246
Quantitative resolution, effect of 'bits', 41
Radius, column, equation for, 19
Recovery of sample, 296,301
drugs, 283
method validation, 291
Refractive index detector, 55,56,71
Resistance to mass transfer, 17
Resolution,
equation for, TLC, 200
peak, 50,64,72
spectral, TLC detection, 224, 225

thin-layer chromatography, 223,225
Response index, 24,32,72
Response of detector, LC, 34
effect on accuracy, 74
Retardation factor, capacity factor,relationship for, 197
Retention indices, 341
Retention process, elucidation of, 397
Robotics, 341
accuracy and precision, 348
Rough hard-sphere theory, diffusion coefficients,
predictions of, 372
Sample application, TLC, 211
Sample concentration
biological samples, 286
gas chromatography 162
water analysis, LC, 76,78
Sample decomposition
drugs, 274
gas chromatography, 159,160
Sample derivatization
biological samples, 303
gas chromatography, 158,159
thin-layer chromatography, 254
Sample detectability, HPTLC, 223,234,229
enhancement of, 254
Sample development, TLC
anticircular, 210
circular, 210
continuous, 201
multiple, 202
two-dimensional, 209
Sample identification
liquid chromatography, 54
methods of, 259
thin-layer chromatography, 233
Sample introduction
automation, 316
gas chromatography, 159
liquid chromatography, 79
reproducibility, 311
Sample mass, LC, 84

Sample molar absorptivity, 236
Sample preparation
automation, 313
drugs, 282
gas chromatography, 157
liquid chromatography, 74
manual, robotic, comparison
reproducibility, 311
robots, 314, 341, 348
Sample recovery, 296, 301
drugs, 283
method validation, 291
Sample size, effect on
extra-column dispersion, 365
peak shape, 383,386
spot broadening, 211
Sample volatility, GC, 161
Scanning densitometry, 220
background correction, 222
computers for, 251
optics, 220
performance, 223
sample identification, 233
Secondary flow phenomenon, 368
Selectivity
chromatographic system, TLC, 200
control of, 3
mobile phase, TLC, 209
spectral, GC, 146
spectral, HPTLC, 235
Sensitivity
detectors, 24
detectors, LC, 32,35
mass, 25
Separation principles, 2
Separation ratio of components, 174
Signal digitization, 41
Signal-to-noise ratio
detection, GC, 111
detection, HPTLC, 223,227
effect on error, 48

Slit dimensions, effect on sensitivity, 225,226,229
Solute
    concentration profile, 380
    critical pair of, 173
    distribution coefficient, 359,376,377,382,383,397
    distribution process, 380
    physico-chemical properties, 359
Solvent control, 82
Solvent migration, TLC, 194
Spot broadening, TLC, 197,205
Spot reconcentration mechanism, TLC, 197,204,205
Standard deviation of peak, 172,173
Static techniques, adsorption isotherm measurement, 391
System noise, LC, 37
Temperature
    control, LC, 80
    effect on injection procedure, GC,160,161
    programming, 12
Thermal conductivity detector, 123
Thin-layer plate
    function of, 23
    performance of, 197
Trace enrichment of sample, 318
Two-dimensional development, TLC, 209,210
Ultraviolet absorption detector, 55,56,71,147,
    219,225,277
Uses of adsorption isotherms, 397
Validation of chromatography methods, 288-297
Van Deemter equation, 15, 66
Variance per unit length, 15
Velocity, optimum, LC, 66
Volatility of sample, GC, 161
Water analysis, LC, 78